The Monumental Challenge of Preservation

Yang Yueluan, "Hebei Shanhaiguan, March 18, 2010," section of
the Great Wall of China. Courtesy of Elmar Seibel, Ars Libri

The Monumental Challenge of Preservation

The Past in a Volatile World

Michèle Valerie Cloonan

The MIT Press
Cambridge, Massachusetts
London, England

© 2018 Massachusetts Institute of Technology

All rights reserved. No part of this book may be reproduced in any form by any electronic or mechanical means (including photocopying, recording, or information storage and retrieval) without permission in writing from the publisher.

This book was set in Stone Serif by Westchester Publishing Services. Printed and bound in the United States of America.

Library of Congress Cataloging-in-Publication Data is available.

ISBN: 978-0-262-03773-0

10 9 8 7 6 5 4 3 2 1

To Sidney and Aaron Berger, and the memory of Raphael (Rafe) Berger

Contents

Illustration Credits

Preface

The premise of this book is that the preservation of the built and natural world is complex and will require new collaborative and interdisciplinary approaches. While this might seem obvious, I selected the word *monumental* partly to suggest the enormous effort needed to preserve the world's tangible and intangible heritage; the book presents cultural, political, technological, economic, and ethical and moral dimensions of preservation. It has been obvious for a long time that the preservation of our heritage cannot rest with preservation professionals alone. We are now at a critical juncture in the world. Civil wars, cybercrime, terrorism, global warming, and decreased government support have put our heritage at perhaps greater risk than at any time since World War II. This moment requires us to think about preservation in the broadest possible contexts.

Throughout this book, I look at particular monuments such as the Bamiyan Buddhas, the Vietnam Memorial, and the Book of Kells to illustrate the many issues that underlie preservation. For example, the Taliban blew up the Bamiyan Buddhas in Afghanistan in 2002; debates are still taking place over whether and how to preserve the remains and also the memory of the Buddhas. When the Vietnam Memorial was still under construction, in the 1980s, visitors started leaving behind symbolic and often ephemeral objects at the site, which the National Park Service continues to preserve, despite the complexity of doing so. Volumes of the Book of Kells have occasionally been lent to other countries under the strong protest of the Irish who believe that this priceless work should never leave the country. In the digital realm, electronic heritage may disappear soon after it is created because not all institutions or countries have the necessary infrastructure in place to preserve it. These are just a few examples of the many dimensions of preservation that this book addresses.

An American example of how contentious the preservation of monuments can be has come to the fore since I wrote this book. It has to do with whether and how to preserve the monuments of the Confederate States of America that broke away from the United States over many issues but primarily over slavery, which they supported and which led to the Civil War. Today, with a divided country and many people supporting hate groups that defend the old Confederacy, the existence of these monuments has caused pain, animosity, and riots. On June 17, 2015, Dylann Roof, a self-proclaimed white supremacist, murdered nine African Americans during a prayer service in the Emanuel African Methodist Episcopal Church in Charleston, South Carolina. Roof later stated that he had hoped to start a race war. On his website he posted images of himself with the Confederate flag, which has become a symbol, for some, of persistent and potentially violent racist hatred. In fact, the flag, which has flown over many public and private buildings in the American South, has been the source of a great deal of contention. Soon after the murders, the South Carolina legislature voted to remove the Confederate flag that flew in front of its State House. On July 10, 2015, the flag was removed. Other towns followed suit by removing their own Confederate flags. Debates ensued over whether all symbols of the Confederacy should be removed.

Opinion polls conducted over the past fifteen years have repeatedly shown that for some white Americans, the Confederate flag represents Southern pride, while for African Americans and others, the flag represents racism.[1] So while the immediate, horrified reaction to the Charleston murders was to remove Confederate flags—and in some cases monuments—two years later, feelings are still divided. But something else has happened. The current U.S. president, Donald Trump, has been tepid in his rebuke of white supremacists and they have felt emboldened. The statues have become *their* symbol. At the recent and violent "Unite the Right" rally in Charlottesville, Virginia, August 11–12, 2017, white supremacists demonstrated on behalf of keeping a statue of Robert E. Lee, president of the Confederate States of America, and himself a symbol of slavery and divisiveness.

Professional associations are calling for thoughtful approaches. The American Historical Association believes that while Confederate monuments should be preserved, historical context is needed.[2] The American Institute for Conservation of Historic and Artistic Works recently issued a "Position Statement on Confederate and Other Historic Public Monuments," in which they state that "while recognizing the right of society to make appropriate and respectful use of cultural property, the conservation professional shall serve as an advocate for the preservation of cultural property." That includes the services of conservators, who can assure the maintenance of monuments, or their safe dismantling and relocation.[3]

Carolyn E. Holmes, a professor of political science at Mississippi State University, believes that Americans can learn from approaches that other countries have taken in coming to terms with tumultuous historical events. South Africa's recent debate about monuments that preserve the memory of white minority rule there is instructive. Whereas in the 1990s, the South African government removed some statues of apartheid-era leaders from city parks and government buildings—and gave them to heritage organizations—today they use a different strategy. The government is constructing new monuments and placing them alongside old ones.[4] Criteria have been developed to determine which of the old ones should continue to be displayed in public places, and which should not—and also how to present the context of the monuments so that they need not engender negative feelings and public displays of anger. We can also learn from Berlin, where there are many monuments that confront the world wars and post-WWII division. This book concludes with a chapter on how the Germans have come to terms with memorializing their own history.

The problems of the past that have affected society with respect to cultures and emotions, laws and desires, greed and hatred, and other such powerful forces are still with us, and will be so forever. How we have dealt with them, and continue to deal with them, are subjects of this book.

I mention legal challenges only in passing, because the subjects covered here are broad; different types of laws pertain to the situations described. Applicable laws may be local, national, or international. Many types of legal issues are identified ranging from copyright to human rights. The full array of laws related to preservation could indeed be the subject of a separate book.

Background

My interest in the many aspects of preservation stems from personal and professional involvement as well as my readings in several fields. In chapter 2 I connect my experiences in Dublin (in the 1970s) and Moscow (in 1991) to my long-time focus on the cultural, political, and social aspects of preservation. Those experiences showed me that conservation and preservation concerns could take place well outside of cultural heritage institutions, in the streets, in war zones, or in public forums. I wrote an essay that explored some of these ideas and it became the catalyst for this book.[5] My awareness of these challenges was further developed after the American invasion of Iraq in 2003. From 2004 to 2011 I worked with colleagues to train Iraqi librarians, archivists, and educators in the Middle East. I have not included that experience in this book; however, in chapter 4 I focus on Syria as only one example of the kinds of problems that now exist in the Middle East.

While my professional activities have broadened my personal perspectives, David Lowenthal broadened my intellectual ones. When I first read *The Past Is a Foreign Country* (recently "revisited" by him)[6] over twenty years ago, a new approach to thinking about preservation opened up to me. I had so focused my attention on the technical and professional literature on conservation and preservation that it never occurred to me that a preservation narrative could be so seamlessly woven into history. It changed the focus of my research. Recently reading Lowenthal's 2015 version of his work, I am newly impressed by his erudition.

Contents

This book is divided into six parts: Context, Cultural Genocide, Approaches to Preservation, Information or Object?, The Greening of Preservation, and Enduring, Ephemeral Preservation. Chapter 1 in part I introduces the book and explores the concept of monumentality in detail. It presents a schema for preservation that relates to the themes in the rest of the book. Chapter 2 presents case studies: the lending of two volumes of the Book of Kells by the Irish for an extended exhibition in the United States in the 1970s, and the taking down in Moscow, in August 1991, of statues of former communist leaders. (The statues were later placed in an area known as "Fallen Monument Park" and "Muzeon Park of Arts" where they still reside.) The studies show how different cultural attitudes play into preservation strategies.

Part II concentrates on cultural genocide. Chapter 3 focuses primarily on one person: Raphael Lemkin, whose groundbreaking work as a jurist during World War II led to the codification of the United Nations Convention on the Prevention and Punishment of the Crime of Genocide in 1948. Lemkin coined the word *genocide*. Chapter 4 looks at a contemporary example of genocide and cultural genocide: the ongoing war in Syria. The chapter provides the historical context for understanding this war and describes current preservation initiatives.

Part III, Approaches to Preservation, looks at two activities that facilitate preservation: collecting and documentation. Chapter 5 highlights some notable collectors who had a strong commitment to preservation—and who collected objects that might have disappeared but for their efforts. Chapter 6 examines the role of Richard Nickel in preserving the work of the architect Louis Sullivan. Most of his papers are at the Art Institute of Chicago, and the architectural fragments that he salvaged from the buildings—in some cases, as they were being torn down—have been distributed (and sometimes preserved) all over the United States.

Part IV, Information or Object?, considers core issues in preserving originals and surrogates. Chapter 7 explores the dynamic between originals and copies and how it affects preservation decision making. Chapter 8 proposes new approaches to preserving digital heritage. Both chapters show some ways in which views about preservation are still evolving.

Part V, The Greening of Preservation, consists of one chapter (9): "Sustainable Preservation." It proposes that the preservation of the built environment shares many issues with that of the natural environment. The chapter begins with John Ruskin, who, in lectures about storm clouds, wrote about pollution, and about the preservation of architecture. The chapter traces some developments in the environmental movement as well as in historic preservation.

Part VI, Enduring, Ephemeral Preservation, explores themes that have been addressed in the first nine chapters, but with a slightly different spin. In chapter 10, I revisit an article that I wrote in 2007 called "The Paradox of Preservation." I reconsider my earlier piece and differentiate between paradoxical and complex situations. Both call for creative strategies. And finally, "Epilogue: Berlin as a City of Reconciliation and Preservation" appraises the ongoing reconstruction of Berlin and the city's approach to creating monuments and memorials. It shows the importance of public engagement in such monumental preservation activities that have been ongoing since the end of World War II. Berlin is something of a preservation laboratory in its efforts to preserve the past while keeping a steady eye on the future.

I have drawn on just a few of many examples that could have been selected to illustrate my points. This book does not attempt to draw lessons and case studies from all parts of the world. Instead, I have focused primarily on Europe, Australia, the United States, and the Middle East. Time and resources have limited me, though there is much to be learned from Africa and from Asia, as well. I have made it a point to include a wide international swath of preservation activities in the journal that I edit, *Preservation, Digital Technology & Culture*. I invite readers to consult it.

The Monumental Challenge of Preservation was written for heritage professionals, students, and scholars in all disciplines that share an interest in preservation. Its goal is to inspire discussions about—and strategies for—preserving our heritage.

To accomplish this, the book touches on a number of complex and multifaceted subjects. No one person can have expertise in all of the subjects covered here. Although I have sometimes stepped outside of my own comfort zone in covering such diverse topics, I hope that *The Monumental Challenge of Preservation* will be a catalyst for wide-ranging discussions and new collaborations.

Michèle Valerie Cloonan
Williamsburg, MA
2017

Acknowledgments

Every book has a cast of supporting players; I am thankful to mine for their assistance. My deepest thanks go to Sidney E. Berger, my helpmeet in every way. He is my biggest supporter, and a wonderful editor. I also owe continuing thanks to my aunt, Marguerite Bouvard, political scientist and human rights advocate extraordinaire. She encouraged (even pushed) me to take courses in political philosophy while I was in college. While I ultimately majored in English, readings by such thinkers as Raymond Aron and John Rawls remained with me, and their link to preservation came to me when I discovered Raphael Lemkin. Linda M. Chan, a former colleague at the Newberry Library—and friend—has provided assistance with almost every research project that I have undertaken. Thanks to Elmar W. Seibel of Ars Libri, a visionary builder of great research collections of art historical materials. He led me to the work of Yang Yueluan.

Thanks also to the following:

Simmons College, School of Library and Information Science
These graduate assistants and students:
Patsy Baudoin, who assisted me in the early phase of the research
Joel DeMelo
Anne Holmer Deschaine
Jennifer Gray
Andra Langoussis Pham
Tricia Patterson
Vanessa Reyes
Erica Ruscio

American Library Association Archives, University Library, University of Illinois
Cara Setsu Bertram

American Library Association
Leonard Kniffel, for the photo "Russians Tearing Down Yakov Sverdlov Monument"

Art Institute of Chicago, Ryerson and Burnham Archives
Nathaniel Parks

Getty Research Institute
Marcia Reed
Mary Sackett

The Grolier Club
Eric J. Holtzenberg

Newberry Library
Linda M. Chan

Simmons College Library
Linda Schuller
Linda Watkins
& the Inter-library loan staff

University Library, University of Illinois
Robert L. Howerton
Daniel Tracy
Sandra L. Wolf

Harvard University Library
András J. Riedlmayer
Jeff Spurr (now retired)
Institute of Museum and Library Services
IMLS and the Salzburg Global Seminar for inviting me to participate in two seminars: "Connecting to the World's Collections: Making the Case for the Conservation and Preservation of Our Cultural Heritage" (2009), and "Libraries and Museums in an Era of Participatory Culture" (2011), which I co-chaired
Mary Chute (now State Librarian of New Jersey)
Nancy E. Rogers (now a consultant)
Marsha L. Semmel (now a consultant)

Salzburg Global Seminar
Susanna Seidl-Fox

MIT Press
Marguerite Avery
Gita Devi Manaktala
Jesus J. Hernandez
Kyle Gipson
Virginia Crossman

The securing of permissions seemed to take almost as long as the writing of this book. My deepest thanks to my friends, friends of friends, family, and colleagues who provided me with images: Sidney E. Berger, Suzy Steele Born, Jonathan Brown, Linda M. Chan, Leonard Kniffel, Kathy Brendel Morse, Anna Okulist, Zaher Omareen, Jane Penner, Barbara G. Preece, Vanessa Reyes, Erica Ruscio, Elmar Seibel, Robert Smith, Bob Thall, and Alexa Zellentin.

Many colleagues have inspired me. In particular, my thanks to Jeannette Bastian, Anne Gilliland, Ross Harvey, and Rachel Salmond.

Finally, thanks to the anonymous reviewers for their insightful suggestions.

I Context

1 Introduction: We Are What We Preserve—
and Don't Preserve

A Monument is a thing erected, made, or written, for a memoriall of some remarkable action, fit to bee transferred to future posterities. And thus generally taken, all religious Foundations, all sumptuous and magnificent Structures, Cities, Townes, Towers, Castles, Pillars, Pyramids, Crosses, Obeliskes, Amphitheaters, Statues, and the like, as well as Tombes and Sepulchres, are called Monuments. Now above all remembrances (by which men have endevoured, even in despight of death to give unto their Fames eternitie) for worthinesse and continuance, bookes, or writings, have ever had the preheminence.

—John Weever, *Ancient Funerall Monuments within the United Monarchie of Great Britain* (1631)[1]

The preservation of cultural heritage is a monumental task. The aim of this book is to explore the preservation of movable, immovable, human-made, and socially constructed heritage in an expansive way. Studies in preservation have tended to reflect particular disciplinary lines: library and archival studies, museum studies, art history, historic preservation, and other fields that focus on preservation from particular vantage points. The premise of this book is that there are broad preservation issues that can be viewed through a variety of disciplines. Law, political science, business, architecture, computer science, geography, cultural tourism and cultural heritage management, sociology, and environmental science (itself a multidisciplinary academic field) are some of the fields that address preservation-related issues. For example, international laws may be essential to recovering stolen artifacts or protecting heritage in war. Natural disasters affect heritage collections; disaster preparedness and recovery can be approached from many angles including public policy, science, and the sociology of disasters. Anthropology and history offer approaches to understanding and addressing multicultural issues in preservation. This book considers many ways of looking at preservation.

In the heritage fields, *preservation, conservation,* and *restoration* are sometimes used interchangeably—and sometimes have distinct meanings. *Preservation* is used most often to refer to the general care of collections and the
built environment. It refers to environmental monitoring, security, conservation, sustainability, and guardianship or stewardship of objects and
collections. With respect to the built environment, preservation activities
may include stabilization, restoration, rehabilitation, or reconstruction.
Conservation usually refers to the treatment and care of objects. The aim of
restoration is to return something damaged or flawed to an earlier known—
or imagined—condition. For Ellen Pearlstein,

> [conservators] are those whose activities are devoted to the preservation of cultural
> heritage. ... Since conservation is also a term applied to the protection of biodiver-
> sity, cultural heritage specialists often distinguish their field by applying the limiting
> phrase "art conservation," perhaps more accurately termed cultural conservation as
> the field encompasses a broad range of specialties including natural history speci-
> mens, archives, and books. ... Restoration has a narrower focus than conservation
> and one that is tied to presentation value. The complete emphasis on restoration
> without regard for preservation or accurate interpretation has been emphatically
> rejected in the modern definition of conservation.[2]

In the 1970s, William Lipe proposed a preventative conservation model
for American archeology in response to the then "growing rate at which
sites are being destroyed by man's activities—construction, vandalism, and
looting of antiquities for the market."[3] This approach included public education, the involvement of archeologists in land use planning, the establish-
ment of archeological preserves, and so on. His work paralleled the efforts
of conservators such as Paul Banks (in libraries and archives) and Caroline
and Sheldon Keck (in museums), as well as others whose conservation work
fostered collaboration, education, and outreach. Now, a generation later,
there are new opportunities for collaboration and outreach.

While "conservation usually refers to the treatment of objects that is
based on scientific principles and professional practices, as well as other
activities that assure collections care and mitigate damage,"[4] its reach continues to expand. Salvador Muñoz Viñas points out that conservation can
also be used in a broad sense to refer to "diffuse boundaries, since it may
involve many different fields with a direct impact on the conservation
object. ... [N]on-conservators' conservation deals with other technicalities
within many varied fields (law, tourism, politics, budget-allocation, social
research, plumbing, vigilance systems, masonry, etc.)."[5] Muñoz Viñas includes

in his book a figure that depicts conservation emerging from a box in which many other fields overlap it.[6]

In the present text *preservation* will be used except in specific instances in which other terms are more precise. My aim here is to concentrate on preservation, with some of the nuances of meaning that the other two terms *conservation* and *restoration* suggest. Preservation continues to evolve. New forces such as social networking are beginning to have a profound influence on the heritage fields. Crowdsourcing is but one example of how the public already is becoming engaged in preservation projects.

While the focus of this book is on preservation, consideration of destruction is a necessary component of that focus. Natural and manmade destruction threatens the preservation of cultural heritage. Natural disasters, ideology (e.g., "cultural cleansing" and cultural genocide), fanaticism, biblioclasm, and military action may lead to the destruction of objects and sites. Administrative decisionmaking for analog and digital collections is yet another cause of destruction; collections may be weeded, records may not be retained, or collections may deteriorate due to neglect. Benign neglect and inaction fall between preservation and destruction. Such inaction may stem from cultural beliefs, the inability to care for collections, forgetfulness, denial of global warming, or any other cause that leads to deterioration or loss.

Figure 1.1 illustrates traditional motivations for preservation as well as behaviors that may have stemmed from historic practices. At the same time, there are emerging trends. Social networking has influenced how we create, disseminate, and share information. Personal information management and personal digital archiving are relatively new disciplines that study how people organize their information. Understanding such information organization practices could lead to the creation of new digital preservation services and products. Community archiving is another movement that is on the rise. Communities may develop new approaches to preserving their records, which may or may not be similar to practices in heritage institutions.

There are four components to the schema shown in figure 1.1: (1) "Motivations," and (2) "Behaviors," which may be based on (3) "Historic Practices," and (4) "Emerging Trends." The areas overlap. Motivations, historical practices, and behaviors may persist at the same time that emerging trends offer new approaches and attitudes toward preservation. For example, while we may continue to preserve our heritage, it may become increasingly difficult to do so as digital art and information are usually diffuse and ephemeral. At

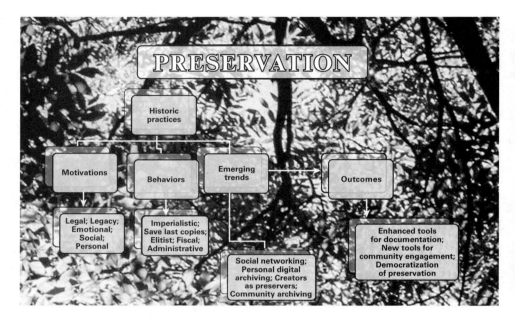

Figure 1.1
Aspects of preservation. Image by Vanessa Reyes and Michèle V. Cloonan

the same time, new technologies may offer ways to "preserve" the old. One such technology, IRENE (Image, Reconstruct, Erase Noise, Etc.), was developed by physicist Carl Haber.[7] (He so named his machine because the first recording that he preserved was the song "Goodnight Irene" by the Weavers.) An old sound recording that is cracked, or otherwise damaged, can now be produced using a combination of optical imaging and high-resolution scanning. IRENE converts the data into patterns that are readable and playable.

The rapid growth of digital personal information—in part brought about through social networking—suggests that all people will have to manage and preserve their own information, or risk losing it. At the same time the preservation of historical sites, analog information, the physical landscape, and digital media is often dependent on legal and social contracts. Thus the social, legal, and political context in which objects reside will impact preservation.

The schema suggests the complex web—and monumental undertaking—of preservation. Thus the inclusion of *monumental* in the title of this book refers to monuments as objects, but it also suggests the monumental efforts of preserving the world's heritage against war, climate change, terrorism, and natural disasters. There are other monumental issues addressed as well, such as community outreach, indigenous heritage rights, sustainability, and what preservation means in an era of increasingly ephemeral heritage. Are *all* texts that are born digital (emails, social media, blogs, tweets, and much more) worth preserving? If so, who will decide what to preserve? In what medium? For how long? And at whose expense?

This book is about preserving cultural heritage writ large, or *monumental preservation*—that is, the aspect of preservation that is broader than the day-to-day practices of conservation and preservation. The *Oxford English Dictionary (OED)* has several definitions of *monument* including (2) "a written document, record; a legal instrument; and (4) anything that by its survival commemorates a person, action, period, or event."[8] Of interest here is that the *OED* definition includes written documents. The idea that the written word can stand as a monument goes back at least two thousand years. Martial observed that "A fig-tree splits Messalla's marble and a rash mule-driver laughs at Crispus' defaced horses. But writings no theft can harm and the passing centuries profit them. These monuments alone cannot die."[9]

Some fifteen hundred years later, Shakespeare echoed Martial's sentiment when he wrote in Sonnet 55 that "Not marble nor the gilded monuments of princes shall outlive this powerful rhyme." For both writers, marble was used as the metaphor for so-called *permanent* monuments.

Similarly, the Latin *monumentum* refers to written memorials, annals, and memoirs.[10] *Monumental* refers to monuments as well as tombs, and it implies a vastness or extensiveness that is far-reaching. Thus by using *monumental*, I am presenting a broad and impressively large and enduring scope of subjects that sometimes directly—and often indirectly—impact preservation. One aim of the book is to inspire the reader to see how these broad themes impact our professional practices—and our lives.

This book also explores the premise that *we are what we preserve*. That statement might seem obvious—a result of the "accidents" of history. In fact, every day we make preservation decisions, individually and collectively. Some decisions may have longer-term ramifications than were originally anticipated. (A collection is discarded, and needed later on. A new road is built near an ancient site, which increases tourism [and traffic], and leads to the deterioration of the site. We choose to delete emails or text messages—or to save them somewhere.)

Finally, a book on preservation must consider why we preserve at all. What are our motivations? What is our motivation discourse—or discourses, as different fields may have distinct reasons for preservation?[11] Any consideration of preservation must be simultaneously broad and nuanced.

What Is a Monument?

In his early definition of monuments (quoted earlier), John Weever (1576–1632) expressed the importance of particular types of public buildings, churches, and memorial structures. Nearly four hundred years later, we have broadened his concept. Monuments may refer to structures that commemorate people, groups, or events. Sometimes structures not intended as monuments become them. For example, the Great Wall of China was erected as a fortification structure, but many see it as a monument to Chinese civilization. At the other end of the scale from the Great Wall, there are small monuments; for example, in California, there is one that was erected to honor the Chinese who built the California railroads (figure 1.2). (However, efforts are afoot to create larger monuments to the Chinese and to resurrect this neglected aspect of California history. For many, a monument must be *monumental*.)

Art historian Alois Riegl (1858–1905) wrote:

In its oldest and most original sense a monument is a work of man erected for the specific purpose of keeping particular human deeds or destinies (or a complex accumulation thereof) alive and present in the consciousness of future generations. It may be a monument either of art or of writing, depending on whether the event to be eternalized is conveyed to the viewer solely through the expressive means of the fine arts or with the aid of inscription; most often both genres are combined in equal measure.[12]

While the aim of Riegl's essay is to examine "artistic and historical monuments," his expansive definition of *monuments* is useful to the aim of this book, which is to show that preservation is a monumental undertaking. He also mentions the importance of inscription, echoing my earlier citation from the *OED* that says that words—written documents—constitute one form of monument.

Sigmund Freud further developed the idea that monuments may evoke "the consciousness of future generations" in his essay "A Disturbance of Memory on the Acropolis."[13] For Freud, a trip with his brother to the Acropolis in 1904 became a metaphor for his own life's journey. He describes his "joyful astonishment at finding myself at that spot," while simultaneously feeling guilt and disbelief that he had made it to Athens. Freud reflected on

Figure 1.2
Chinese Railroad Workers Memorial Monument, Gold Run, CA. Photo by Sidney E.
Berger and Michèle V. Cloonan, February 12, 2017

what he and his brother had accomplished as compared to their father, who had neither gone to university nor traveled very far. Freud experienced these epiphanies on the site that is a powerful symbol of Western civilization—and Greek culture and mythology. The essay was written in 1936 as a letter to Romain Rolland "on the occasion of Rolland's seventieth birthday" and toward the end of Freud's life. In the thirty-two years between Freud's visit to the Acropolis and his letter to Rolland, this monumental experience still had a hold on his sensibilities.

Freud conveys the power of his visit this way: "When, finally, on the afternoon after our arrival, I stood on the Acropolis and cast my eyes around upon the landscape, a surprising thought suddenly entered my mind: 'So all this really *does* exist, just as we learnt at school!'" At the end of the essay he corrects himself—he never "doubted the real existence of Athens. I only doubted whether I should ever see Athens."[14]

Freud's multilayered interpretation of his experience at the Acropolis is suggestive of the experiences that many of us have when visiting famous landmarks. I remember the range of emotions I felt when I visited the Great Sphinx of Giza: first, excitement at seeing it, followed by disappointment that it seemed much smaller "in real life" than in the photos that I had seen of it. I felt frustrated that the site was surrounded by vendors aggressively trying to sell their souvenirs, uncomfortable in the extreme heat of that day, and cross that there was nothing to drink—only souvenirs for sale. At the same time I felt lucky to have been able to travel to Egypt, and guilty for feeling cross. Yet I was more taken by the dogs sleeping in the small amount of shade provided by the tourist buses—how they got there, how they would survive—than in the Sphinx that I had traveled so far to see (figures 1.3 and 1.4).

The Sphinx was less monumental than I had anticipated; and all these years later I still remember vividly the responses I had then. So the experience—accompanied by my memory—had its own "monumental" impact on me. This is one of the messages of the present book: that *monumental preservation* can be an intensely personal phenomenon, at the same time that it is a social or historical phenomenon.

Our complex and evolving relationship with monuments—and how (or whether) to preserve them—is a recurrent theme of this book. For some people, all monuments are worthy of preservation. For others—as we shall see in my discussion of the Bamiyan Buddhas—some monuments must be destroyed. For Weever in particular, a monument implied "eternitie" while

Figure 1.3
The Great Sphinx of Giza, Egypt. Photo by Michèle V. Cloonan, 1980

Figure 1.4
Tour buses and dogs near the Great Sphinx of Giza. Photo by Michèle V. Cloonan, 1980

Riegl allowed for the more temporal "consciousness of the future." How might we define *eternity* or *future* today? Our time horizons for the preservation of monuments are now much shorter, and the concept of *monument* continues to evolve. While this book does not focus on individual monuments, they appear throughout the book and different insights about them are presented.

In *Learning from Las Vegas*, Denise Scott Brown, Robert Venturi, and Steven Izenour debunk the idea that buildings needed to be "heroic."[15] They contend that "learning from the existing landscape is a way of being revolutionary for an architect."[16] The architecture of the desert has a lot to teach us about the role of the natural landscape in the West. But most striking, perhaps, is the synthetic aspect of Las Vegas: commercialism, popular culture, superhighways, and the automobile that came to characterize this mid-twentieth-century type of urban environment, propelled, in part, by President Dwight D. Eisenhower's Federal-Aid Highway Act of 1956 that created a 41,000-mile national system of interstate highways. The photographs that Brown took of Las Vegas are stunning, exemplifying the theme of architecture and communication that runs through the book, and demonstrating a new type of monumental beauty—though not everyone would view Las Vegas that way. But perhaps it is not too much of a stretch to contend that there is something monumental about the architecture that is visible from so many miles away. And sometimes—depending on the weather conditions and the time of day—the buildings are almost mirage-like, as one approaches them on Interstate 15.

These photographs illustrate the monumentally garish and the monumentally beautiful aspects of Las Vegas (figures 1.5 and 1.6). The image of the Stardust Hotel against a magnificent sunset is a dreamscape, while the Strip

Figure 1.5
Minor commercial buildings and signs on The Strip, 1965, by Denise Scott Brown.
Permission of MIT Press

Figure 1.6
Stardust Hotel and Casino, 1965, by Denise Scott Brown. Permission of MIT Press

Figure 1.7
Vietnam Wall with "offerings." Permission of Kathy Brendel Morse, photographer

is an assault to the senses. The city can have the same impact on people as more traditional monuments have: awe, wonder, and even, for many, horror. Monuments may engender a visceral or emotional response in those who encounter them, and not all responses will be positive.

Another structure that led to a new way of thinking about monuments, and one exemplifying this emotional response, is Maya Lin's Vietnam Memorial (dedicated in 1982)[17] (figure 1.7). An open competition was held to select the design. Four criteria were included in the call for submissions: the design must include the names of everyone who was killed or missing, harmonize with the National Mall site, facilitate a healing process, and not

Figure 1.7 (continued)

make a political statement.[18] Lin submitted her bold, modern design while she was an undergraduate student at Yale University. Instead of a monument consisting of idealized warriors, soldiers on horseback, or laurel wreaths, Lin designed a stark, simple wall onto which would be etched the names of the 58,272 Americans who had died or disappeared in Vietnam. This moving work of polished black granite is shaped like an elongated V—evoking Vietnam, veterans, victims.[19] Part of the power of Lin's creation—beyond the names—is the tactility of the monument, which allows people to touch the letters and not be distracted by anything pictorial.

The design was controversial, inspiring a change in the United States in the way that war memorials were conceptualized. Vietnam was not a "heroic" war and thus called for a new approach to memorialization. The monument gives voice to those who died in the war rather than presenting an idealized view of a solitary, abstract figure. Yet the initial reaction to Lin's design was fierce, referred to as a "black gash of shame" among many other derogatory descriptions by veterans, politicians, and critics alike. Lin's design unleashed the many complexities of the Vietnam War and Americans' response to it. The emotional response I just spoke of, then, can be negative as well as positive.

To appease the many critics, the Vietnam Veterans Memorial Fund added a traditional bronze statue in 1984 titled "Three Servicemen" by the sculptor Frederick Hart. This piece was criticized for being sentimental, for detracting from the Maya Lin design, and for ignoring the women who served in the military. This led to the addition of another bronze representational monument, Glenna Goodacre's "Vietnam Women's Memorial," in 1993. Her statue depicts three nurses who are attending to a wounded male soldier. The nurse in the foreground and the soldier in this nouveau Pietà resemble the Virgin Mary and Christ.

The statues are positioned in awkward juxtaposition to "the Wall," as Lin's monument is often called. Some feel that the integrity of her design was compromised by the addition of the Hart and Goodacre statues. And yet their inclusion provides an ongoing dialogue about the meaning of monuments. Lin's work is one of the most visited monuments in Washington, DC; in 2011, it was the second-most visited after the Lincoln Memorial.[20]

One important aspect of the Vietnam Veterans Memorial has preservation implications: people began to leave personal objects in front of the memorial even before it was completed. In this respect, Lin created an early interactive experience with mourners that is similar to the instant memorials that are

created alongside highways, at accident scenes, or at any locale where someone has met an untimely death. Tributes that have been left at the Wall are
gathered by the National Park Service and stored in a climate-controlled facility in Latham, Maryland. To date, some four hundred thousand items are in
the collection. A virtual exhibit, "Items Left at the Wall," a collection of 500
objects, was released online on August 6, 2015.[21] There are plans to add more
items to it, which will turn this online resource into another monument.

Judith Dupré believes that

the rise of the spontaneous memorial—the practice of leaving mementos at locations
where tragedy has occurred—is directly related to the ubiquity of photography and
reaches back still further to the cartomaniacal blurring of the distinction between
the individual and the celebrity....Photography abets the spontaneous memorial
itself by reproducing its image, which inspires yet another segment of the population to make a therapeutic deposit at the growing mountain of offerings.[22]

Today memorializing (a result of monument construction) also takes place
online, through a variety of websites that make it possible to commemorate
the deceased. These include 1000 Memories, GoneTooSoon, Remembered
.com, and Memorial Matters. This is but one example of the role that social
networking plays in preservation activities; others will be described in later
chapters. It is probable that in time, even more memorializing will take place
online. However, the creation of physical monuments will not abate soon.
People still seek physical "places of memory."

A visual debunking of the heroic monument is Michael Elmgreen and
Ingar Dragset's "Powerless Structure, Fig. 101" seen here on the Fourth
Plinth of Trafalgar Square in 2012–2013 (figure 1.8).[23] The northwest plinth
was originally intended to have on it an equestrian statue of William IV,
but the funding never materialized. Since the late 1990s artists have been
invited to create works for the plinth, which remain on display from a year
to a year and a half. Many of the works playfully challenge the notion of
the traditional monumental statue. Elmgreen and Dragset suggest in their
statue that there is a heroism to childhood that is worthy of commemoration. But there are many possible readings of the work, including the
obvious play on important men on horses. If monuments aim to elicit emotions, then humor and playfulness are not out of place, as the Elmgreen
and Dragset statue shows. (Their powerful "Memorial to the Homosexuals
Persecuted under the National Socialist Regime" is discussed in this book's
"Epilogue: Berlin as a City of Reconciliation and Preservation.")

Figure 1.8
"Powerless Structure," by Michael Elmgreen and Ingar Dragset. From Flickr Creative Commons

There are also nontraditional monuments such as the AIDS Memorial Quilt that was begun in Atlanta, Georgia, in 1987. It is monumental in ambition and size. By the end of 1987 there were 1,920 panels; today there are 98,000, with new panels being created all the time. (There are some "half a million articles in the Quilt Archives which include letters, cards, photos, books, clippings, artwork, etc. that we receive with the panels."[24]) Dupré describes the AIDS Memorial Quilt as "an atypical monument. It is the quintessential antimonument, not heroically ascendant, but laid low on the ground. It is soft, domestic, intimate yet epic. Made of acres of fabric … [it is] more vulnerable than stone or steel, it is a democratic monument. …"[25] Because quilts can

Figure 1.9
Panels from the AIDS Memorial Quilt, n.d. Permission of the NAMES Project Foundation

be made by anyone, and with salvaged scraps just as easily as with expensive fabrics, they are an egalitarian medium. The quilt (see figure 1.9) is maintained and exhibited by the NAMES Project Foundation.[26]

While the quilt may not have some of the characteristics of "fixed" monuments (e.g., composed of enduring materials like stone or metals), it has its own monumentality in its size and in the number of people who have contributed to it—and also the fact that it is not fixed in time and place, like a traditional monument, but is like a living entity that continues to grow. At the same time, it cannot be preserved in the traditional sense because its composition continues to change. Like performance art, it may only be possible to preserve versions of it with scanned or photographic images as it appeared in particular times and places.

* * *

What role does preservation play when objects are at the point of destruction, or have been destroyed? Architectural historian Richard Nickel's comprehensive documentation of buildings slated for—or in some cases undergoing—demolition is the subject of chapter 6, "Worth Dying For? Richard Nickel and Historic Preservation in Chicago." Nickel died in 1972 when architectural gems were being demolished at an astonishing rate in large American cities, often in the name of urban "renewal." His efforts to preserve the buildings of Dankmar Adler and Louis Sullivan were seldom successful, though he was able to salvage pieces of some of their buildings. And his careful research on and photo-documentation of Sullivan's work after the Sullivan/Adler partnership dissolved have provided invaluable resources to scholars. A magnum opus on Adler & Sullivan, which Nickel started but never finished, was completed by Aaron Siskind, John Vinci, and Ward Miller and published in 2010.[27]

The dynamics between preservation and destruction have been playing out powerfully with the Buddhas of Bamiyan, Afghanistan. The two standing Buddhas were constructed in the sixth century C.E. along the Silk Road, where there were once many Buddhist monasteries. In 2001 the statues were destroyed by the Taliban who had expressed their intention to destroy the "infidel" monuments beginning in the late 1990s. The world community tried to convince Taliban leaders not to destroy them, but those efforts were not successful, and the Buddhas were dynamited over a period of about two weeks in March 2001. The niches in which the Buddhas once stood remain. Behind the niches are wall paintings, and evidence suggests that a third Buddha is still buried (figures 1.10 and 1.11).

UNESCO inscribed the "Cultural Landscape and Archaeological Remains of the Bamiyan Valley" a World Heritage Site in 2003. Since then, a UNESCO-funded project is sorting the fragments. Claudio Margottini, an engineering geologist for the conservation of cultural heritage, has edited a book on rehabilitating the site.[28] The possibility of reconstructing at least one of the statues with the remaining fragments is under consideration, though some people have expressed doubt as to whether the statues should be rebuilt.[29] (For an update, see chapter 10, Preservation: Enduring or Ephemeral?")

Michael Falser, a research fellow and the chair of Global Art History at Heidelberg University in Germany, has suggested that the Taliban were protesting as much the globalizing concept of "cultural heritage" as they were the Buddhas themselves.[30] In other words, an international approach to preservation may threaten regional hegemony or religious beliefs. In this sense, groups such as the Taliban and ISIS (the Islamic State of Iraq and Syria), also referred to as IS and ISIL (the Islamic State of Iraq and the Levant), are taking

Figure 1.10
"De Provincie Bamian." Image from *De Aardbol: Magazijn van heden daagsche land-en volkenkunde* 9 (1839). Permission of Europeana

Figure 1.11
Bamiyan Buddha after March 11, 2001. Permission of Europeana

back what they see as "theirs" (all kinds of human-made objects and architectural sites); their view is that they own these things, which they believe gives them the right to destroy them. (This idea will be considered in more detail in chapter 4, "Documenting Cultural Heritage in Syria.")

A Japanese-American artist, Hiro Yamagata, came up with another approach to recreating the site. A decade ago he designed a laser project ("Laser Buddhas") that could be powered by windmills and solar panels. His aim was to "reproduce" the statues in holograms at the site. Images of his laser creations can be found on the web. While this project never came to fruition in Afghanistan, Yamagata has exhibited his installation.[31] More recently, a Chinese couple, Janson Yu and Liyan Hu, have created a 3D Buddha laser projection light show to re-create the two Buddhas; images of their creation are available on the web.[32] They received permission from the Afghan government and UNESCO to display the Buddhas in situ for one night: June 7, 2015.[33] (New digital technologies are making it possible for tourists to "visit" historic sites that for preservation reasons are not open to the public. Such projects are described in chapter 7, "What Are We *Really* Trying to Preserve: The Original or the Copy?") This goes beyond preservation to reimagining.

As the examples in this chapter show, there are as many attitudes and responses to preservation as there are cultures, governments, and political movements. An example of large-scale cultural genocide took place in Bosnia during the Balkan Wars of the 1990s with the destruction of the Bosnian National and University Library in Sarajevo, as well as other archives, libraries, and sites in the country.[34] The large-scale destruction of monuments is taking place now in Iraq and Syria, mostly at the hands of ISIS. However, there is a new twist. ISIS doesn't destroy all the antiquities that it steals: it sells many of them on the black market. The money supports the ISIS army. Therefore, it profits from the sales of small antiquities while destroying monumental structures that are too big to sell. The destruction makes for great PR for their cause, and ISIS seems to carefully record its plundering. Moreover, ISIS will kill anyone who attempts to thwart these objects' destruction. Since 2014, conservators, lawyers, guards, curators, and arts administrators have been murdered in Iraq and Syria.[35] Most recently, in August 2015, Islamic State militants beheaded the archeologist Khaled al-Asaad, an elderly scholar, in Palmyra, Syria. He apparently refused to disclose where valuable artifacts had been moved for safekeeping.[36] Some monuments are simply beyond saving.

* * *

Strategies for preserving cultural heritage continue to evolve and are described throughout this book. Many professional organizations, foundations, and government agencies contribute to global preservation efforts. Yet new collaborative alliances are needed as well. Environmental historian Marcus Hall has called for "more conversations between restorationists of nature and of culture."[37] Among the professions that share the broad concerns elucidated in the definitions given at the beginning of this chapter, scholars in these fields do not necessarily communicate across professional domains. This is due in part to the fact that those with an interest in preservation—the term that I privilege in this book—work in different academic disciplines, attend different professional conferences, and publish in different venues. While there are ample examples of interdisciplinary work across these fields, this book seeks to open the door wider than it is now.

Two additional terms that are much used today need to be defined here: *cultural heritage* and *cultural memory*. What do they signify? John H. Stubbs, a historic preservationist, defines *cultural heritage* as

1. The entire corpus of material signs—either artistic or symbolic—handed on by the past to each culture and, therefore, to the whole of humankind. ... 2. The present manifestation of the human past. ... 3. Cultural Heritage possesses "historical, archaeological, architectural, technological, aesthetic, scientific, spiritual, social, tradition or other special cultural significance, associated with human activity." 4 "Evidence of Cultural Value" at a site comes from "the comparative quality of a mixture of different factors: construction; aesthetics; usage/associations; context; and present condition."[38]

Cultural heritage can also be defined more broadly to include the preservation of the tangible (monuments, buildings, works of art, books, documents) and the intangible (customs, beliefs, lore, unrecorded music, language). It refers to inheritance and legacy as well as to traditions. Cultural heritage emanates from particular societies at particular times, and is passed from one generation to the next. But much gets lost, and what is not lost will always be vulnerable to the vicissitudes of man and nature.[39]

Cultural memory is another term that is often used in association with preservation. It is the process of recalling what we have learned—individually and collectively. Philosophers have speculated about individual memory for some two thousand years.[40] However, in the twentieth century, in the wake of World War I, art historian Aby Warburg and sociologist Maurice Halbwachs independently came up with ideas of a collective memory or social memory. Memory cannot be separated from the conditions in which it is formed—and later recalled. Jan Assmann further developed ideas about how collective knowledge evolved from socialization and customs. Eventually scholars referred to this scholarship as cultural memory. As Robert DeHart puts it, "…cultural memory represents a reconstructed past that serves the present needs of a group. … A group's cultural memory is reinforced and transmitted to other individuals in forms such as commemorations, rituals, monuments, oral traditions, museum exhibits, books, and films."[41]

The social process of memory is studied by scholars in many fields including history, literature, art history, psychology, cognitive sciences, computer science, and library, archives, and information studies. The preservation of heritage facilitates memory.

In *Realms of Memory*, Pierre Nora describes external sites of memory such as museums and monuments that can preserve historic links between people and historic events.[42] However, the representation of past events is necessarily interpretive. Many have questioned the accuracy of particular

Figure 1.12
Section of the Great Wall of China, Hebei Shanhaiguan, January 28, 2013, by Yang Yueluan. Courtesy of Elmar Seibel, Ars Libri

interpretations. One example is Colonial Williamsburg (in Virginia), which has been subject to reinterpretation since its opening in the 1930s. Criticisms have included its commercialism, its neat-and-tidy facade, and its inaccurate portrayal of the African-American experience. Museum exhibitions have occasionally been subject to protests over "how we remember." Perhaps the most famous episode in recent decades concerned the Enola Gay exhibit at the Smithsonian Institution, which had been meant to commemorate the fiftieth anniversary of the end of World War II. (The Enola Gay was the B-29 airplane that dropped the atomic bomb over Japan.) The initial exhibit was canceled as a debate ensued about how the museum should represent the history of the dropping of the bomb.

Such debates force us to consider what and how we preserve for the long term, not just the short term. While Michael Elmgreen and Ingar Dragset play with the notion of "short term" in their "Powerless Structure," figure 1.8, above, Yang Yueluan shows how the ruins of the Great Wall of China persist in the long term, but in new ways (figure 1.12). This book will consider the short- and long-term aspects of preservation.

2   A Tale of Monuments in Two Cities

I can trace the roots of this book to two unrelated but impactful early events in my professional life: interning in the conservation lab at Trinity College Library in 1976–1977 while preparation was under way for the controversial *Treasures of Early Irish Art* exhibition that traveled to and through the United States from 1977 to 1979; and attending an international professional conference in Moscow in late August 1991 as the putsch took place, and where I witnessed a crowd take down statues of early communist leaders. These events also underscore my use of "monumental" in the title of this book.

The usefulness of these examples is threefold: (1) they occurred in the last quarter of the twentieth century, which makes it possible to take some measure of what has happened since then; (2) as I had direct experience in each event, I have had some three decades to reflect on each; and finally, (3) recent activities in Ireland, Russia, and Ukraine invite us to reflect once again on 1976 and 1991—and to see that some events that relate to heritage may never be "resolved."

In the Beginning...

In early 1976 the Minister for Education in Ireland, Richard Burke, announced that there would be an international traveling exhibition of Ireland's greatest treasures, monuments of Irish history that included the Book of Kells (figure 2.1), the Ardagh Chalice, and the Tara Brooch. According to the article in the *Irish Times* that reported this news, the exhibition *Treasures of Early Irish Art* was in part "an effort to put forward 'a happier and truer image of Ireland than the one too often propagated at the present time.' [Burke] hopes it will 'persuade as large a public as possible to look beyond our contemporary troubles to the record of a rich and ancient civilization.'"[1] In

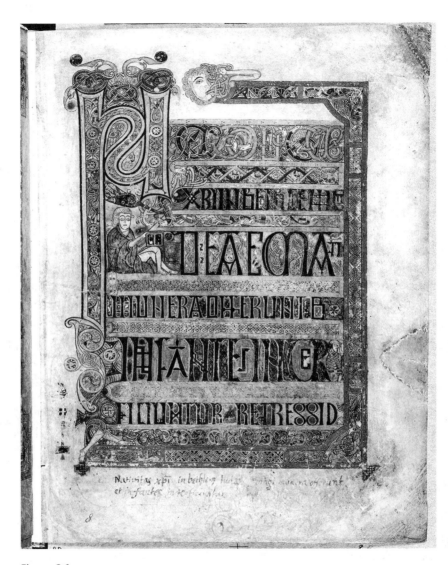

Figure 2.1
"In the Beginning," the Book of Kells, MS58 fol. 8r, 800 C.E. Permission of the Board
of Trinity College Dublin

Ireland the news was greeted by academics and the general public alike with concern. Opinions about the proposed exhibition, as well as updates about its preparation and, later, its travel and reception in the United States, were regularly reported on by the *Irish Times* throughout 1976 and 1977. Later, in 1978, fresh criticisms were leveled when *Treasures of Early Irish Art* was reviewed in the *New York Review of Books*.

One of the exhibition's several controversies[2] surrounded the fragility of some of the objects to be loaned, especially the Book of Kells. One Irish scholar has described the manuscript as a "national monument of the kind that everyone has heard of, and of whose nature everyone is, however vaguely, aware."[3] The show was to travel to five American museums[4] over two years (1977–1979)—a grueling schedule. Were the potential risks worth the perceived public relations benefits? The director of conservation at Trinity College Library at the time, Anthony Cains, was opposed to the lending of even one volume of the manuscript; two of the four volumes were ultimately exhibited. The complete manuscript had left Ireland only once: to undergo conservation treatment by Roger Powell in Britain in 1953. (The first volume, "Matthew," was exhibited at the Royal Academy in 1961, "in aid of a Library extension appeal."[5]) In addition to the risk of loss that could result from such travel, Cains and others worried that exposing the manuscript to completely different environmental climates from that in the UK and Ireland could cause permanent damage to the vellum or the pigments or both. When Cains lost the battle to prevent the Book of Kells from traveling, he insisted that a world expert on conservation standards for artifacts in transit and on exhibition, Nathan Stolow, be brought in as a consultant. Stolow was well known for his research on and experience with the packing and transport of artifacts for exhibition. He was also an expert on the environmental design of exhibition cases. By the 1970s, he had already published a well-respected book, *Controlled Environment for Works of Art in Transit*.[6] He would publish much more on exhibitions and other conservation subjects over the next several decades.[7] Conservators at New York City's Metropolitan Museum of Art, in consultation with Cains and Stolow, designed exhibition cases that could maintain a continuous, controlled environment for the manuscripts.

In bringing in an international expert, Cains used a strategy that others who are mentioned in this book in subsequent chapters have also used: wielding the influence that you have. Cains did not succeed in keeping the Book of Kells in Ireland, but he was able to insist that the highest conservation standards were adhered to in the exhibition preparation. Similarly, as we will see in chapter 6, Richard Nickel came to understand that even when he could not save a particular landmark building, he could convince Chicago's mayor's office to pay for complete photo documentation of and research on the particular building that was to be torn down—*when there was enough time.*

Not everyone was opposed to the exhibition of Irish treasures; some even advocated for it, maintaining that if only risk factors were weighed, few items would ever be lent for exhibition. In a similar vein, Dr. Maureen de Paor, who wrote a chapter in the accompanying exhibition catalog, *Treasures of Irish Art, 1500 BC–1500 AD,*[8] pointed out "that we could not reasonably expect other countries to send works on loan unless we were prepared to loan some ourselves." Dr. de Paor lamented the fact that in exhibitions of Celtic art exhibited abroad in the past, Ireland had been only scantily represented, and she said she was convinced that "a great deal of care and trouble had been undergone in order to provide for the safety of this exhibition."[9]

In a July 8, 1977, column, de Paor's husband, Liam de Paor, a regular writer for the *Irish Times* and himself a contributor to the *Treasures* catalog, further heralded the benefit of having Irish artifacts widely exhibited. He cited the example of the successful *Treasures of Tutankhamun* exhibitions.[10] An *Irish Times* reader soon after shot back that "Dr. Liam de Paor...suggests that it is a good idea that Ireland's treasures should be sent on a travelling exhibition. It is a deplorable idea. Dr. de Paor cites for his case the successful Egyptian exhibition in London. But there is no comparison between the wealth and abundance of Egyptian antiquities and the few treasures now left in Ireland. It is not necessary to send our treasures on a trip which may cause possible loss or deterioration."[11]

Other objections to Ireland's treasures being lent overseas were voiced in the *Irish Times*. For example, visitors to Ireland during that time might be "disappointed at finding none of the major early art treasures in our museums"[12] if so many irreplaceable objects were sent out of Ireland at once.

A little additional controversy ensued in New York City. A prominent scholar of early medieval art, David H. Wright, published an essay in the *New York Review of Books* in which he criticized the show as not only "conceived in secrecy and born in controversy,"[13] but also poorly displayed ("Truly the Dark Ages revisited, and indeed at times it also seemed like the Black Hole of Calcutta")[14] and accompanied by a catalog whose "quality is mixed."[15] In fact, Wright had nothing positive to say about the catalog of which he criticizes the scholarship as well as the photography. A rather anemic response to Wright's review was launched by David Greene, a former president of the Royal Irish Academy, and published a couple of months after Wright's review.[16] Rather than refute Wright's points, he merely states the importance of early Irish art, and thus the exhibition.

And yet, despite the controversies, the *Treasures of Early Irish Art* exhibition became a blockbuster show, drawing enormous crowds across the United States. The exhibition cases for the Book of Kells and the *Book of Durrow* were modified for permanent use at Trinity College Library. The college created a new exhibit area for the Book of Kells, which, before the traveling exhibition, was perched almost inconspicuously at one end the Library's Long Room (figures 2.2 and 2.3). Moreover, the United States exhibitions paved the way for later traveling shows in Europe and Australia in the early 1980s, and later. (A complete list of the shows is included in the entry for the Book of Kells in the UNESCO Memory of the World Register.) After volume two, "The Gospel of Mark," was lent to the National Gallery of Australia, the Board of Trinity College decided not to lend the Book of Kells again.[17] But the Book of Kells is such a major tourist attraction today that there are almost always lines to see it—which was not the case in the 1970s (figure 2.4). Certainly, any work as monumental as the Book of Kells will be the subject of new debates in the future.

"Lenin, Kaput!"

In contrast with the Irish treasures is the case of the "fallen statues" in Moscow, which has been playing out since August 1991 during the putsch to take control of the country from Soviet president Mikhail Gorbachev.[18] Demonstrators used cherry pickers to pull down statues of prominent communist-era leaders (figure 2.5). There was no intention of destroying the statues. Rather, their removal was a symbolic action to signify that a new era was beginning in what was then still the Soviet Union. As of this writing, there still is no permanent home for the statues, though they now reside in a park. Indeed, the statues have been caught up in Russian and Soviet identity politics. Also, perhaps, they have played a role in the drama of remembering and forgetting in Soviet and Russian politics.

Figure 2.2
The Book of Kells was formerly displayed in a less prominent area, n.d. Permission of
the Board of Trinity College Dublin

Figure 2.3
Part of today's exhibition area, n.d. Permission of the Board of Trinity College Dublin

Figure 2.4
The Book of Kells is a major tourist attraction, 2017. Permission of Alexa Zellentin, photographer

In Russia (as in the former Soviet Union), one can find many instances of national and political identity coupled with objects of commemoration. For example, in August 1991, I was in Moscow to attend the International Federation of Library Associations and Institutions (IFLA) conference. At the many libraries and archives that IFLA participants visited, we saw examples of deteriorating collections. But the most memorable experience for me was watching with my colleague, the late Susan G. Swartzburg, the statues of early Bolshevik leaders Felix Dzerzhinsky and Yakov Sverdlov being taken down by a crowd using a cherry picker. The fallen leaders were then hauled off by trucks to an unknown destination. What would happen to the statues? Would they be preserved? When we left Moscow a week later, we could not determine their fate.

A partial answer can be found on numerous web postings. Fallen Monument Park was created in 1992. Located outside the Krymsky Val building in Moscow, the park previously has been known as Park of the Fallen Heroes, Park of Totalitarian Art, and Park of the Arts.

Figure 2.5
Russians tearing down Yakov Sverdlov Monument, 1991. Permission of Leonard Kniffel, photographer

According to a website in 2011:

In October 1991, when the Soviet Union collapsed, smaller socialist realism statues of Soviet leaders and unidentifiable *workers and peasants* were removed from their pedestals, hauled to the park and left in their fallen form. They were rectified later, although missing original pedestals. In the 1990s these statues shaped the park outline, but as more and more modern sculpture was added and as the trees grew up, they became a less obvious minority. Opening animation in the film *Golden Eye* and one of the levels in the *Ninetendo64* [*sic*] game of the same name were based on images of the fallen monuments, although in both the game and film the park was located in Saint Petersburg.

In 1995, Muzeon added a World War II section—these sculptures, of the same socialist realism vintage, were never displayed in open air before. In 1998 the park acquired 300 sculptures of victims of communist rule made by Evgeny Chubarov, installed as a single group. The park also holds temporary summer shows of modern artists.[19]

The narrative on the website is inaccurate. The first statues were taken down in August, not October, 1991. Susan Swartzburg and I each kept diaries from August 17–26, 1991. Mine records that the statue of Felix Dzerzhinsky was pulled down from Sverdlov Square on the evening of Thursday, August 23, and the Yakov Sverdlov statue was removed the next day. At 11:30 p.m. that night a Russian man approached us yelling, "Lenin, Kaput!" In other words, the statue of Lenin would be next.[20] We did not witness the removal of any more statues, but our diary entries suggest that these activities would continue on or after August 24. More fundamentally, the web entry does not address the issue of what it has meant to preserve these sculptures. Does this itself signify anything? Apathy? Or is 1991 now as remote as World War II?

That the statues were moved to a park rather than destroyed makes them worthy of study and reflection. And that "they were hauled to the park and left in their fallen form" is worthy of interpretation. Was their placement meant simply as a metaphor for the downfall of the Soviet regime?

The excerpt that follows was copied from the web, on February 10, 2015:

Fallen Monument Park (formerly called the Park of the Fallen Heroes) is a park outside the Krymsky Val building in Moscow shared by the modern art division of Tretyakov Gallery and Central House of Artists. It is located between the Park Kultury and the Oktyabrskaya underground stations.

The origins of this expatriate English name [sic] unknown; in Russian, the park is either simply named Sculpture Park (Russian: Парк Искусств, Park Iskustv) of the Central House of Artists (Russian: Парк скульптуры ЦДХ) or referred to by its legal title, Muzeon Park of Arts.

See figures 2.6–2.8. The shifting web texts make it difficult to establish a consistent chronological narrative.

The future of Fallen Monument Park is uncertain. Yelena Baturina, an entrepreneur, had planned to demolish the gallery building and erect a Norman Foster-designed *Orange*—a mixed-use project spanning the current territory of the park. The project was canceled, presumably by former President Dmitri Medvedev. Baturina currently lives in London,[21] having moved there after her husband, Yury Luzhkov, mayor of Moscow from 1992 to 2010, was forced to step down by Medvedev. Foster's project is not going forward.[22]

What is revelatory in the web narratives is how the events of 1991 have been conflated with other historical periods so that the toppled monuments have become a subset of "socialist realism statues of Soviet leaders." The survival of the statues, for now, indicates a cultural position of great significance: monuments are monuments, and their destruction may have huge implications. Their survival is telling. Also of significance in their current placement is that the curators of Fallen Monument Park have broadened the collection to include modern sculptures; thus the earlier statues "became a less obvious minority." Now the park—and possibly its statues—are threatened. According to the online travel guide Atlas Obscura, the land may be

Figure 2.6
Statues of Joseph Stalin and Vladimir Lenin. From Flickr Creative Commons

Figure 2.7
Muzeon Park of Arts, 2011. Permission of Anna Okulist, photographer

Figure 2.8
Vandalized Stalin. From Flickr Creative Commons

redeveloped.[23] Will the loss of the park result in the loss of the statues and the memory of August 1991?

Isn't memory always at risk of disappearing? The short answer is "yes," and the many reasons for such loss are considered throughout this book. The fate of the statues is reflected in Soviet—and now Russian—collective memory.

The fallen statues are but a window into a more complex social and political landscape that exists in every country. The preservation cannot be complete of any monument that is taken out of its political, social, aesthetic, and/or religious context. This point is important. The statues, where they were originally located, had meaning—in fact, more than meaning: *significance*. Part of that significance was in the power of their sculptural form, and part was in where they were—prominently displayed in a public place. Removing them from that place and essentially hiding them away in a park, practically buried among many other pieces, seriously dilutes their meaning, their power, their monumental significance.

In 1966, the All-Russian Society for the Preservation of Historic and Cultural Monuments was founded, a result of pressure at the time by Russian nationalistic intellectuals. They were protesting Khrushchev's closures and destruction of churches between 1959 and 1964, as well as the extensive rebuilding of Soviet cities, such as Moscow, which led to the tearing down of old buildings, parks, and squares.[24] In another piece, Hosking writes that in Soviet societies, "The party leaders are the self-appointed custodians of history. That is the fundamental grounding of their claim to legitimacy."[25]

Benjamin Forest and Juliet Johnson, writing on Soviet-era monuments in Moscow, delineate three possible fates for them: co-opted/glorified, disavowed, or contested.[26] They describe these as follows: "Co-opted/Glorified monuments are maintained or further exalted. Disavowed sites are literally or symbolically erased from the landscape either through active destruction or through neglect by the state. Contested monuments remain the objects of political conflict, neither clearly glorified nor disavowed."[27]

The statues of the early Soviet leaders might fall into all three categories. First they were co-opted—removed from their pedestals and taken to a park for fallen heroes. As time has passed, they have perhaps become disavowed as their place in the park is becoming marginalized—or manipulated. Depending on the fate of the park, the statues may become contested. Should they continue to be preserved? These categories are not necessarily mutually exclusive: the role of the monuments may shift as Soviet history is reinterpreted or as Russian identity changes.

Forest and Johnson noted that the meaning of the monuments has changed with their physical context: "these Soviet icons moved from public spaces representing national identity, to a literal trash heap, to a tourist attraction, and finally to a historical and artistic display."[28] Their article was published in 2002; if the information on the website for the park[29] is accurate, and its future is uncertain, the meaning and context of the monuments may change again.

If the statues in Fallen Monument Park disappear or are scattered, and they *are* remembered, it may be because of their depictions in a James Bond movie or a Nintendo game. Such popular cultural juxtapositions suggest, perhaps, an unintentional cultural bricolage—the combining and recombining of artifacts that result in something new.[30] However, in the process, their place in the historical record will change.

Leninfall

Soon after Russia invaded Ukraine in 2014, statues of Lenin were vandalized, dragged through the streets, or even destroyed in Kiev (and in other Ukrainian cities). Vladimir Sorokin compared the "Leninfall" to the statues that were more politely removed in Moscow in August 1991. He points out that the Lenin statues were felled

during the brutal confrontation on Kiev's Maidan Nezalezhnosti (Independence Square), when Victor Yanukovych's power also collapsed, demonstrating that a genuine anti-Soviet revolution had finally occurred in Ukraine. No real revolution has happened in Russia. Lenin, Stalin, and their bloody associates still repose on Red Square, and hundreds of statues still stand, not only on Russia's squares and plazas, but in the minds of its citizens.[31]

Sorokin is further of the opinion that Boris Yeltsin's revolution was "velvet" because it did not bury the Soviet past and did not pass judgment on Soviet crimes. According to Justinian Jampol, however, it will take more than the smashing of Lenin statues to change the course of Ukraine:[32] "In Ukraine, the struggle for ownership of the past continues to unfold. And while the protesters' destruction of monuments might hasten their victory, it does not mean they will have an easy time putting their fragmented country back together."[33]

Meanwhile, in Russia, president Vladimir Putin has recently promoted the preservation of Russia's architectural heritage following the claim by Konstantin Mikhailov, of the preservation group Arknadzor, that "unchecked construction was threatening the country's U[NESCO]-listed sites." Mikhailov holds the president responsible for construction around Zaryadye Park, "a project proposed by Putin to replace a demolished hotel near Moscow's Red Square."[34,35] Galina Malanicheva, chair of the central council of the All-Russian Society for the Preservation of Historical and Cultural Monuments, has stated that ten to fifteen nationally registered Russian monuments are demolished each year; in 2015 alone, more than eighty monuments were lost. Because of a complicated listing system in Russia, only 10 percent of Russia's monuments are listed.[36] Putin's newfound enthusiasm for preservation may or may not be permanent.

The Soviet statues bring home two points. First, preserving monuments is clearly more than a physical process or practical measure; it may be powerfully political. There is a fiscal component, as well as emotional and spatial ones, and the practicability of the situation must be addressed: are there people with the expertise and the clout to make preservation happen? But all the expertise, motivation, money, and good intentions to preserve our objects of culture may have no sway in the face of political pressure and power. Ultimately, if Vladimir Putin decrees that all monuments "must go" they will go, regardless of the oratory such an edict will engender from the world (and as considered in chapter 4, "Documenting Cultural Heritage in Syria").

Second, as Sorokin has suggested, if such drastic measures are taken and monuments must "go," their preservation may be only as photographs or in memories. And memories are finite. Even the transmission of memories from generation to generation—even with a photographic record (for memories come with people's interpretation)—dilutes meaning. In the face of political power, monuments are not always preservable and their significance can wane when monuments *are* saved but their contexts deteriorate.

Troubles Behind, Troubles Ahead?

In Dublin, the four volumes of the Book of Kells have not left Ireland since 2000. However, a recent news item is reminiscent of Richard Burke's 1976 reference to "The Troubles." The Alfred Beit Foundation decided in 2015 to auction several paintings that once hung in Russborough House. Maintenance of the historical house is expensive, and sale of the paintings will cover the deficit. The last two people to own the home, Sir Alfred Beit and Lady Beit, were British. In 1974, they were beaten and tied up by the Irish Republican Army (IRA), and a number of their paintings were stolen. The paintings were recovered and the couple donated most of them to the National Gallery of Dublin.[37,38] Reminiscent of the press for the *Treasures of Early Irish Art* exhibition, there is great opposition by some to the sale of the paintings. The public dialogue continues.

Selling the Beit paintings could be seen—in an oblique way—as a form of destruction or loss of cultural heritage. The paintings were *significant* in their context. They would have been removed from that context, diluting their cultural significance; and they could have wound up anywhere (a truly private and thus invisible collection? A remote museum?). And as those in the preservation field (and in libraries, archives, and museums) have proclaimed for generations: selling off one's treasures to cover costs of infrastructures is foolish. Important items disappear, the money disappears, the entity that did the selling may survive but with seriously reduced cultural information, and our cultural heritage is diluted. Whatever monuments we have—in the loss of their property—are diminished. Any kind of such dispersal makes our monuments less monumental and flies in the face of preservation.

As a result of my experiences in Dublin and Moscow, I witnessed conservation and preservation concerns playing themselves out on a world stage, embodying the issues that engaged (or enraged) the public. The examples I have given in this chapter illustrate how preservation may be a catalyst for deeper social, cultural, political, and ethical issues. These issues may play out for decades or even centuries. In Ireland, there were strong sensitivities about a traveling exhibition that would remove the country's treasures for two years. There was a sentiment in some quarters that Ireland had already been looted of much of its heritage. If the plane carrying the items crashed, too many irreplaceable treasures would be lost, a perspective born of Ireland's history.

In Moscow, in August 1991, preservation came to imply something quite different. A successful coup meant pushing the communist past aside. Statues of early Soviet communist leaders standing in visible public spaces were intolerable to the reformers. Soviet-era legal documents promulgated in 1948 and 1973 sought to identify, maintain, and protect historic and cultural monuments.[39] As I have noted, it is significant that protesters took down *but did not destroy* the statues.

Figure 2.9
"Good-bye Party Comrades!" by Antanas Sutkus, 1991. Gelatin silver print. Permission of the Art Institute of Chicago

My final Moscow diary entry records: "As we were driving out of Moscow, a group of people was pulling out the base under the statue of Dzerzhinsky that was removed Thursday night. (8/24/91)"[40] Susan Swartzburg had heard that the statues would be placed in "a park for fallen Soviet leaders." According to an entry in her diary, there was even talk that Lenin's tomb on Red Square would be moved to the country.[41]

In Moscow, the urge to remove the symbols of political oppression from sight was not trumped by the urge to preserve. Rather, the survival of the statues more likely signifies the lack of tidy political or social resolutions. Antanas Slutkus's 1991 photograph "Good-bye Party Comrade!" (figure 2.9) illustrates a lack of "tidiness" as we see the top part of the statue has been plucked off its legs.

As I have shown in this chapter, monumental preservation has political and emotional components and often the best intentions in the world, facing the political realities, may not be enough to effect true preservation.

In the following chapters, we will see a variety of monumental forces at work on preservation.

II   Cultural Genocide

3 To Destroy Culture: Raphael Lemkin's Lessons about Genocide and How They Relate to the Preservation of Cultural Heritage

Cultural heritage has been created and destroyed with regularity around the globe throughout history. There are many causes of destruction, among which are war, politics, imperialism, greed, power, prejudice, and natural or human-made disasters. Sometimes the destruction of cultural heritage is the byproduct of war, and sometimes violence against human beings that has nothing to do with war is accompanied by the deliberate destruction of their cultural records.[1] This chapter traces the concepts of cultural genocide and ethnic cleansing. While these two concepts differ somewhat, they refer to the intentional destruction of the culture of a particular group of people. In one culture's desire to wipe out an enemy—or to destroy any race or group of people—members of that culture may decide (for a variety of reasons) that it is in their best interest to do it *completely* if possible in such a way that no trace of those who are vanquished is left behind. Such complete annihilation, first of all, hides the destruction and thus yields a world in which there is no blame for such destruction since no one would ever know there was an annihilation if no traces at all were left of the annihilated. Further, even a small trace of the destroyed people could be a seed that could grow into a reborn culture—or a movement that would condemn the perpetrators of the genocide for their actions.

Complete removal of traces of any peoples is practically impossible, however. But for every destroying body, wiping out other cultures will always be condemned by most who see cultural loss as humanity's loss. "Any man's death diminishes me," John Donne wrote.[2] We are all the poorer for the loss of any culture (regardless of how brutal or objectionable it may have been) for the *cultural* elements—there is always something worthy of preservation even if only as a negative example that we can learn from.

Let me be clear at the outset: genocide is evil while cultural genocide is pernicious. To force people to relinquish their language, beliefs, dress, or other aspects of their culture lessens the dignity of people, and is a soulless act. This chapter considers the roots of genocide and cultural genocide.

Preservation efforts may sometimes take place at "the nexus between cultural heritage and human rights," as András J. Riedlmayer has described the destruction of libraries and archives during and after the Balkan Wars of the 1990s in the former Yugoslavia.[3] As I have said, a major cause of human rights abuse is war. A brief overview follows of how human rights and cultural heritage fit into the context of war. The chapter then looks at some international organizations that were created before and after World War II, as well as the many declarations and charters that have been enacted in support of human rights, generally, and cultural heritage, specifically. Recent approaches to human rights will be touched on. The chapter concludes with an analysis of gains that have been made over the past seventy years as well as the profound challenges that continue to effect the preservation of cultural heritage. Finally, I suggest some new preservation strategies.

There are countless histories of the destruction of cultural heritage in particular times and places. Some scholars have attempted to chronicle the history of such destruction. For example, Raphael Lemkin wrote an unfinished history of genocide that included the killing of groups of people and the destruction of their cultural heritage.[4] More recently, Fernando Báez wrote a history of the destruction of books from ancient Sumer to modern-day Iraq.[5] Rebecca Knuth documented regime-sponsored destruction of books in *Libricide*.[6] Unfortunately, such sources cannot possibly convey the many complex political and social circumstances that led to the destruction of heritage objects. Further, it is not always possible to know when and how the destruction occurred. *A thorough understanding of the causes of destruction of cultural heritage as well as the limitations of what can be done to prevent it may lead to effective strategies for its preservation.* And even if we do understand the when and the how, it will often be difficult for us to know fully *why* it happened—and why it was possible to allow the destruction in the first place.

The connection between cultural heritage and human rights is recent. Indeed, the modern concept of human rights is fairly new, even though the ideals behind "rights" and "liberty" can be found in ancient writings. The United Nations—established in 1945 at the end of World War II—established a Human Rights Commission. With Eleanor Roosevelt as its chair, the

Figure 3.1
Eleanor Roosevelt and the UDHR. Courtesy of the Franklin D. Roosevelt Library and Museum

Commission drafted the Universal Declaration of Human Rights (UDHR) in 1948 (figure 3.1). Two accompanying covenants elaborated on the meaning of the UDHR: the International Covenant on Economic, Social and Cultural Rights, and the International Covenant on Civil and Political Rights, though these were not finalized until the 1970s. The three documents are known as the International Bill of Human Rights.

According to one source, the term *human rights* came into use in the late eighteenth or early nineteenth century.[7] However the notion of rights—for some if not all people—goes back to the Magna Carta. Modern conceptions of human rights date back to the Enlightenment. Yet despite the UDHR, realization of universal human rights continues to be elusive.

Likewise, treatises on the conduct of war and rules for its engagement have existed for thousands of years—and in works that continue to be read today, such as Sun Tzu's *The Art of War*.[8] In that as well as other early works such as the Indian *Book of Manu*, the Old Testament, and classical Greek and

Roman literature are the seeds of the idea that people have individual rights. The humane treatment of captured soldiers and civilians, and women, children, and the elderly in particular, is often touched on in these works. But how can such behavior be monitored? Who has legal or moral authority to judge those who do not observe the rules of war? And don't the spoils of war always go to the victors, who usually call their own shots? Judging those who violate the rules of war is one thing. Another is stopping them—or bringing them to justice after the fact. Who has the authority *and the power* to do so?

A shift in societal approaches to war emerged in the nineteenth century. With the rise of the sovereignty of nation states, war became a "legitimate" tool with which nations could pursue national policy objectives.[9] At the same time, massive casualties and the inhumane treatment of prisoners of war as well as of noncombatants caught the attention of many outsiders, including journalists, doctors and nurses on battlefields, social reformers, and other observers. One person, Geneva businessman and reformer Henri Dunant,[10] was an organizer of the First Geneva Conference for the Amelioration of the Condition of the Wounded Armies in the Field (1863–1864), and in the same year he was one of the founders of the Red Cross.[11] Even today there is tension between the notion of state sovereignty and human rights laws that seek to limit how countries treat their citizens and their captives.

New approaches to thinking about armed conflict ensued. Across the ocean, also in 1863, German-American jurist Francis Lieber prepared for the Union Army General Orders No. 100: Instructions for the Government of Armies of the United States in the Field (The Lieber Code), which set out rules for the humane treatment of civilian populations in areas of conflict and forbade the execution of prisoners of war. The Lieber Code further sought to ensure that "classical works of art, scientific collections, or precious instruments, such as astronomical telescopes, as well as hospitals, must be secured against all avoidable injury, even when they are contained in fortified places while besieged or bombarded (Article 35)."[12] The ideas expressed in Article 35 of the Lieber Code were further developed in The Hague international peace conferences of 1899 and 1907 whose principles would underpin the "Hague Conventions," and later the Convention for the Protection of Cultural Property in the Event of Armed Conflict (1954; and the 1999 Second Protocol).[13]

The 1935 Treaty on the Protection of Artistic and Scientific Institutions and Historic Monuments, also known as The Washington Pact and the Roerich Pact, was named after Nicholas Roerich (1874–1947), a Russian artist who lived for a time in the United States; he believed that there was a need for international legal recognition of the importance of cultural objects, and that the protection of objects should take precedence over any military necessity. While the 1935 Treaty was signed by twenty-one North and South American states, it was later overshadowed by the 1954 Hague Convention.[14] Significantly, the convention recognizes that there is sometimes a military necessity for destroying "cultural property"—the term used in the convention. For example, historic structures may be situated in the only possible line of fire. The Roerich Pact is recognized in the preamble to the 1954 Hague Convention, which itself was created as an expansion of the earlier Hague documents, and in response to the damage inflicted upon "cultural property" during World War II. It describes the responsibilities of the occupying powers in protecting cultural property.

It is perhaps unfortunate that the 1954 Hague Convention selected the term *cultural property* because it implies that there is always clear ownership of cultural records—which there is not. Nicholas Roerich's "artistic and scientific institutions and historic monuments" strikes a more neutral tone, but his words were never adopted. Over the past two decades, however, it has become increasingly common to use the phrase *cultural heritage*, not *cultural property*. Using *heritage* rather than *property* conveys the idea that in one sense, heritage belongs to everyone and *heritage* is a more encompassing notion than is *property*.

The Hague definition is as follows:

Article 1. Definition of cultural property

For the purposes of the present Convention, the term "cultural property" shall cover, irrespective of origin or ownership:

(a) movable or immovable property of great importance to the cultural heritage of every people, such as monuments of architecture, art or history, whether religious or secular; archaeological sites; groups of buildings which, as a whole, are of historical or artistic interest; works of art; manuscripts, books and other objects of artistic, historical or archaeological interest; as well as scientific collections and important collections of books or archives or of reproductions of the property defined above;

(b) buildings whose main and effective purpose is to preserve or exhibit the movable cultural property defined in sub-paragraph (a) such as museums, large libraries and depositories of archives, and refuges intended to shelter, in the event of armed conflict, the movable cultural property defined in sub-paragraph (a);

(c) centers containing a large amount of cultural property as defined in sub-paragraphs (a) and (b), to be known as "centers containing monuments."

Despite the progress that the early Hague and Geneva conventions represented, two events of the early twentieth century—the Armenian Massacre (Medz Yeghan) in 1915 and World War I, the so-called "war to end all wars"—made it clear that laws of war are easily broken. For example, the Hague Conventions of 1899 and 1907 forbade the use of "poison or poisoned weapons," yet these were widely used in World War I. Thus the Geneva Protocol for the Prohibition of the Use in War of Asphyxiating, Poisonous or other Gases, and of Bacteriological Methods of Warfare of 1925 was added to the Hague Conventions of 1899 and 1907. It banned all forms of chemical and biological warfare. As recent history shows—for example, in Iraq and Syria—all the laws and conventions in the world will not stop a country from ignoring these prohibitions.

The Kellogg-Briand Pact (or Pact of Paris) of 1928, officially the General Treaty for Renunciation of War as an Instrument of National Policy, called for the peaceful settlement of disputes. Signatory states promised not to use war to resolve conflicts that could be settled by other means. It became the international legal basis for the prosecution of those responsible for waging wars.[15] Unfortunately, the Kellogg-Briand Pact did nothing to prevent World War II or conflicts that immediately followed the pact, such as Japan's invasion of Manchuria in 1931. With respect to this pact, the historian of the U.S. Department of State said, "its legacy remains as a statement of the idealism expressed by advocates for peace in the interwar period."[16] Despite its limitations, the pact became the legal basis for the notion of crime against peace.

Yet international strategies that can mitigate the effects of war are critical to a more peaceful world. (How effective they are is still debatable, but we must have them nonetheless.) International humanitarian law (IHL) seeks to balance humanitarian concerns with military necessity. It evolved from the two components of the laws of war, *jus ad bellum* (reasons a state may engage in war—criteria for a just war) and *jus in bello* (laws that come into effect once the war has begun—how wars should be fought). The Geneva Conventions (1863, 1906, 1929, and 1949) and the Hague Regulations and Conventions (1899 and 1907, 1954, and the 1999 second protocol to 1954) are examples of IHL, as is the International Court of Justice, which is part of the United Nations.

The International Criminal Court (ICC), which was established in 1998, like the International Court of Justice, is seated in The Hague. It investigates

Figure 3.2
The International Court of Justice, Old Building, 2014. Permission of Robert Smith, photographer

claims of and hears cases on international crimes of genocide, crimes against humanity, and war crimes (figure 3.2). While relatively recent, it was planned for several decades after World War II, but was delayed in part by the Cold War. The ICC draws on the UN Convention on the Prevention and Punishment of the Crime of Genocide of 1948 (hereafter referred to as the Genocide Convention, or UNGC[17]) and considers cases referred to it by the UN International Court of Justice. (To date twenty-two cases and nine situations have been brought before the ICC.[18])

Despite the various treaties that were signed after World War I, many abuses of international law occurred in the lead up to and during World War II. Many provisions in the Geneva and Hague conventions were violated. At the same time, some people were trying to develop strategies to avoid such violations. One of them was a Polish-Jewish jurist and philologist, Raphael Lemkin (1900–1959; see figure 3.3), who spent many years thinking and writing about discrimination, killing, and crimes against humanity.[19] He and his family lived in a part of Poland that was absorbed by Russia during the partitions, and they suffered many indignities due to restrictions forced upon

Figure 3.3
Raphael Lemkin, back row on the far right. Permission of the United Nations Photo Library

Jews; they lived through several pogroms. Later as the Germans invaded Russia during World War I, the Lemkin family home would be destroyed. Perhaps as a consequence of his background, Lemkin chose to spend the rest of his life writing about genocide—a term that he coined—and applying the principles of international law to end it. Indeed, according to Thomas de Waal, Lemkin sought "to get the concept of genocide enshrined in international law."[20]

As a young law student at Lvov University, Lemkin became interested in Turkey's massacre of the Armenians and the question of who should bear responsibility for those crimes. He was also disturbed by the destruction of Armenian art and culture—and what that loss would mean to the larger Ottoman Empire. So strong for him was the link between cultural and physical deprivation that the cultural aspect of human massacre would become part of his definition of *genocide*. During the time the atrocities were committed against the Armenians, *crimes against humanity* was the term used. It originated in the 1899 and 1907 versions of the Hague Convention.

Lemkin practiced law in Poland and participated in the activities of the International Association of Penal Law and other such groups throughout the 1930s. He also began promoting his own ideas. When Lemkin first started conceptualizing terms to describe mass killings and the destruction of art and culture, he proposed adding two new crimes covered by international law: *barbarity* (defined as "the extermination of ethnic, social, and religious groups by means of massacres, pogroms or economic discrimination") and *vandalism* ("the destruction of cultural or artistic works which embodied the genius of a specific people").[21] In a 1933 report to the Fifth International Conference for the Unification of penal law, sponsored by the League of Nations, he promoted the terms. He hoped that they could be part of "an international treaty … declaring that attacks upon national, religious and ethnic groups should be made international crimes, and that perpetrators of such crimes should not only be liable to trial in their own countries but, in the event of escape, could also be tried in the place of refuge, or else extradited to the country where the crime was committed."[22] He was not successful.

Perhaps it was just as well that Lemkin's proposal was rejected; use of those terms might have led to ambiguity and might not have been strong enough to have moral—or legal—force. Over the next several years Lemkin continued to write and to attend conferences, and to make professional connections that would eventually help him to escape to the United States in 1941. Through one of his many contacts he was hired to teach at Duke University law school. Tragically, while he was able to get out of Europe, most of the members of his family in Poland were murdered; he had not been able to convince them to leave.

It was only a couple of years after he moved to the United States that Lemkin came up with a concise term to describe the large-scale massacre of groups of people. As noted earlier in this chapter, *genocide* is the term he introduced. Lemkin devoted chapter 9 to the concept in his 1944 book *Axis Rule in Occupied Europe: Laws of Occupation, Analysis of Government, Proposals for Redress.*[23] Based on the Greek word *genos* (race, tribe) and the Latin *caedere* (killing), he meant for *genocide* to correspond to such words as *homicide, infanticide, fratricide,* and so on. "Genocide is a new word," he wrote, "but the evil it describes is old. It is as old as mankind."[24]

Axis Rule in Occupied Europe, which was commissioned and published by the Carnegie Endowment for International Peace in Washington, D.C., identified eight components of genocide. Later, Lemkin reduced it to three when he was campaigning for the 1948 Genocide Convention: physical, biological, and cultural. Lemkin, careful legal scholar that he was, documented all the information that he could gather about how the Germans imposed their edicts and laws on the countries that they occupied, and the ways in which they destroyed national order and customs while imposing Nazi culture. He also demonstrated that the Germans had created an elaborate pecking order. These were the seeds of Nazi genocide. Countries that were "more German" were treated less harshly. Finally, Lemkin recorded the laws against Jews that were enacted in the various occupied countries and regions. (For example, denial of wages to Jews, the requirement that they wear insignia, sequestration of their property, and so on.) It is particularly poignant to realize that while writing the book, Lemkin was trying to convince Americans of the atrocities that were taking place in Europe. And he had yet to learn of the fate of members of his family.

Lemkin did not believe that *genocide* had to refer to the immediate mass killings of all members of a nation. ("Nation" could refer to "families of minds" and a nation could exist "within the minds of people."[25]) Rather the term could refer to the coordinated actions that would eventually lead to the "disintegration of political and social institutions, of culture, language, national feelings, religion, and the economic existence of national groups, and the destruction of the personal security, liberty, health, dignity, and even the lives of the individuals belonging to such groups."[26] Today, the term *cultural genocide* is used to refer to "culture, language, national feelings," but for Lemkin, *genocide* incorporated everything listed above, and thus for him use of the term *cultural genocide* would have been redundant.[27]

After publication of *Axis Rule in Occupied Europe,* Lemkin played supporting roles in the Nuremberg Trials (though his concept of peacetime genocide was not used in the judgment delivered against the accused[28]), and, as I noted earlier, in the codification of the UN's Convention on the Prevention and Punishment of the Crime of Genocide of 1948.

Lemkin was unsuccessful in his attempts to include the cultural aspects of genocide in the convention. Some countries, including, notably, Australia, Sweden, the United States, and Canada, felt that the notion of the destruction of customs and culture was too vague and could be too widely applied. *Possibly these countries also feared that their own actions against indigenous peoples would be conceived of as genocide under Lemkin's definition.* Thus they rejected Lemkin's multifaceted conception and thought that use of the term *genocide* should be limited to cases of mass murder. Despite Lemkin's persistent lobbying for the Genocide Convention in New York and in Washington, D.C., the United States did not ratify it until the late 1980s. Article 2 of the convention condemned

any of the following acts committed with intent to destroy, in whole or in part, a national, ethnical, racial or religious group, as such:

(a) Killing members of the group;
(b) Causing serious bodily or mental harm to members of the group;
(c) Deliberately inflicting on the group conditions of life calculated to bring about its physical destruction in whole or in part;
(d) Imposing measures intended to prevent births within the group;
(e) Forcibly transferring children of the group to another group.[29]

While Lemkin was immediately recognized for his coinage of *genocide*, the rest of his work was nearly forgotten after his death. In the 1990s, however, Lemkin's work received renewed attention and *Axis Rule in Occupied Europe* was reissued in 2004. In 2005 and 2013, issues of the *Journal of Genocide* were devoted to him.[30] (The fact that there are journals with such titles as *Journal of Genocide* and *Genocide Studies and Prevention* points to Lemkin's legacy.) Today's scholars examine Lemkin's work in new interdisciplinary contexts. There is a strong focus on cultural genocide. In addition to legal scholars, historians, political scientists, and sociologists all study Lemkin.

Immediately after World War II, there was a strong conviction that the world could not ignore the mistreatment and suffering of peoples. The United Nations had been founded in 1945, as World War II was drawing to a close. (Since its founding it has consisted of a General Assembly, a Security Council, an Economic and Social Council, the UN Secretariat, and the International Court of Justice.) By 1949 four international documents had been created that continue to hold force: The United Nations Charter, The United Nations Declaration of Human Rights, The Geneva Conventions, and the Genocide Convention. As already discussed, none of these documents adequately addressed cultural genocide.

Since World War II, there have been many conflicts and wars around the globe. The wars and insurgencies in the former Yugoslavia from 1991 to 2001 (War in Slovenia, Croatian War of Independence, Bosnian War, Kosovo War, Insurgency in the Presevo Valley, and Insurgency in the Republic of Macedonia) provide unfortunate examples of ethnic cleansing, genocide, and cultural genocide. All of the conventions and international laws cannot stop an oppressive group from trying to wipe out its "enemies" and their culture. Unfortunately, laws and conventions have little or no power when an aggressor wants to do something "forbidden." If the act they want to do violates a law, nothing is going to make them stop. The law is merely an inconvenience easily ignored. So, while conventions and laws against genocide and "cultural cleansing" have their place, it is discouraging to see how they can be dismissed or ignored by an aggressor. A civil society must do all it can to stop human rights abuses.

The term *ethnic cleansing* is often used to refer specifically to the wars in the former Yugoslavia. *Ethnic cleansing* is the "purposeful policy designed by one ethnic or religious group to remove by violent and terror-inspiring means the civilian population of another ethnic or religious group from certain geographical areas."[31] Ethnic cleansing includes deportation so it does not necessarily include murder. Therefore it is seen by some as being distinct from *genocide*, in which mass murder is ubiquitous. In genocide and ethnic cleansing, targeted groups are generally ethnic, religious, or indigenous. During the Balkan wars, ethnic cleansing and genocide were practiced; for example, in Srebrenica, Muslims were driven from their homes and some eight thousand men and boys were murdered. (In the *Radislav Krstić* case, the International Criminal Tribunal for the former Yugoslavia [2004] found that genocide had been committed in Srebrenica in 1995; in 2007 the International Court of Justice upheld that finding.)[32]

It is instructive to analyze the term *ethnic cleansing*. It implies that something is dirty, infected, or even dangerously filthy, and the act of cleansing is actually doing good. It is akin, in its thinking, to the term *manifest destiny*, used by American settlers in the New World to justify killing millions of Native Americans, or stealing their land and forcing them to live in the worst places in the country, when the settlers wanted their lands.

To add to their justification that their destiny was obvious (manifest, and thus almost an obligation)—and with the attitude of great self-righteousness—settlers called the indigenous peoples *savages* to prove that savage things needed to be cleansed. They were not displacing or killing human beings—women, men, and children—they were cleansing the land of savages, and making the world cleaner and better. The point is, those doing the aggression—those perpetrating the genocides—have ways of justifying their crimes to make them seem logical and beneficial acts.

Still, there is no consensus as to what constitutes a massacre, a genocide, or ethnic cleansing. As Jacques Semelin has pointed out, "not every massacre can be considered genocide and genocide is composed of one or more massacres."[33] Yet definitions are sometimes too wide or too narrow. According to Israel Charny, any massacre constitutes genocide.[34] The opposite opinion is held by Steven Katz who believes that there has been only one genocide—and it was against the Jews.[35] Semelin has tried to bridge these differences by creating the *Online Encyclopedia of Mass Violence*, which is interdisciplinary in approach. It intentionally uses the more neutral term *mass violence* although it does include scholarship on genocide.[36] Nonetheless, the term *mass violence* is a weakening of the concept of *genocide* since the latter has "killing" embedded into its etymology while *violence* merely implies vicious actions, the "exertion of any physical force so as to injure or abuse."[37] *Violence* does not even necessarily lead to killing.

Semelin identifies what he perceives as a "UN school of genocide scholars." They believe that the UNCG offers a legal definition of genocide that has standing. But it is important to understand how genocide is different from other forms of mass murder. According to Meghna Manaktala, "in the final reckoning, the actual issue with regards [sic] to genocide, as with any other violation of human rights, is that of ensuring that it is prevented, or if not, tackled as soon as possible to bring to justice those responsible for such an appalling crime. Any definition is only as good as the contribution it can potentially make to this prime concern."[38]

Large-scale cultural genocide took place during the wars in the former Yugoslavia. András Riedlmayer and Andrew Herscher have meticulously documented the destruction of libraries, archives, churches, and other heritage sites in Sarajevo, Kosovo, and many other locations in the former Yugoslavia.[39] The importance of the documentation of cultural heritage sites is a theme that will be considered in detail in the following chapters. In this instance, the documentation was used as evidence in the courts of law that tried war criminals like Slobodon Miloševic, former president of Serbia and president of the Federal Republic of Yugoslavia. I must stress here that the documentation of cultural heritage implicitly acknowledges the slaughter of people. The things the victims created—churches, books, documents, and so on—prove that they existed. This is preservation in its fullest sense.

While cultural genocide is not yet recognized in international law, its relationship to genocide was acknowledged by the International Criminal Tribunal for the former Yugoslavia in the *Radislav Krstic* case. The Trial Chamber wrote that where there is physical or biological destruction there are often simultaneous attacks on the cultural and religious property and symbols of the religious group as well, attacks which may legitimately be considered as evidence of intent to destroy the group. In this case, the Trial Chamber will thus take into account as evidence of intent to destroy the group the deliberate destruction of mosques and houses belonging to members of the group.[40]

The concept of cultural genocide has found resonance among indigenous peoples. Shamiran Mako explains,

cultural genocide has often been invoked as a conceptual framework for the non-physical destruction of a group.... The non-physical destruction facet of genocide, which Lemkin emphasized as part of his original use of the term, is a fundamental factor for assessing the cultural destruction of a group because it exposes other categories of group destruction that are often overshadowed by the limited definition of the Genocide Convention.[41]

Not all human rights violations relate to war. Forcing native peoples to assimilate into a dominant culture was one way of committing cultural genocide. Examples of such practices abound in American, Latin American, Canadian, and Australian history, to name just a few places. It is no wonder that these dominant cultures were slow to recognize the criminal nature of this subjugation.

Mako traces post-World War II attempts to acknowledge the rights of indigenous peoples, beginning with the International Labour Organization's Indigenous and Tribal Populations Convention C107 of 1957. However, this document was limited in scope and later there was a move to address indigenous rights comprehensively in the United Nations, beginning with the 1982 Working Group on Indigenous Populations (WGIP), that addressed such issues as genocide, human rights, cultural preservation, environmental protection, and social and economic development.[42]

The WGIP was abolished in 2006 and replaced by the UN's Expert Mechanism on the Rights of Indigenous Peoples (EMRIP). A UN declaration had been in the works since the 1990s. Article 7 of the EMRIP draft had included cultural genocide. But just as objections had been raised in 1948 to the notion of cultural genocide, or to be more precise, Raphael Lemkin's full definition of genocide, objections were raised again—by some of the same countries. Thus, in the United Nations Declaration on the Rights of Indigenous Peoples (UNDRIP) (2007), the term *cultural genocide* does not appear. This demonstrates the influence that major world powers have, trying to protect their own reputations in the face of their own *barbaric* acts—to return to Lemkin's notion of *barbarity* and *vandalism*.

The UNDRIP contains much language about cultural rights, beginning in the Annex, in which it is stated that indigenous peoples may "freely pursue their economic, social and cultural development."[43] The protection of indigenous cultural heritage is addressed in Articles 3, 7, 8, 11, 12, and 31. For example, Article 11 covers the protection of traditions and customs as well as archeological and historical sites, artifacts, designs, and ceremonies. Article 31 is explicit about these rights:

Indigenous peoples have the right to maintain, control, protect and develop their cultural heritage, traditional knowledge and traditional cultural expressions, as well as the manifestations of their sciences, technologies and cultures, including human and genetic resources, seeds, medicines, knowledge of the properties of fauna and flora, oral traditions, literatures, designs, sports and traditional games and visual and performing arts. They also have the right to maintain, control, protect and develop their intellectual property over such cultural heritage, traditional knowledge, and traditional cultural expressions.

Thus the UNDRIP emphasizes the rights of indigenous or native peoples and the obligations of countries to protect those rights. Realizing the mechanisms for assuring such rights is the ultimate challenge. Will the exclusion of cultural genocide from this declaration be an impediment? Moral persuasion is no substitute for international customary law. And just because UNDRIP exists, how can its mandates be policed and enforced in the face of a government bent on the destruction of its "enemies"? Enforcement in the face of outright, blatant violations of the declaration is often (even usually) impossible.

The United Nations Educational, Scientific and Cultural Organization (UNESCO) was founded in 1945. (Plans for it were made in 1942 as countries began to prepare to reconstruct their educational institutions after the War.[44]) Many of its activities have related directly to preservation, including the 1972 Convention Concerning the Protection of the World Cultural and Natural Heritage. This resulted in a list of sites that were inscribed beginning in 1978. The 1972 convention was enhanced in 1992 with the creation of the Memory of the World Programme to protect irreplaceable library treasures and archive collections.

There are critics of UNESCO's approach. Its selection of heritage sites—and of library and archival "treasures"—is highly selective and political;

thus it does not advance a systematic approach to preservation. Schüller-Zwierlein points out that "the discourse [of 'cultural treasures']...does not define what is to happen to those objects that are *not* defined as treasures."[45] Preservation must be ongoing and systematic. And the discourse for preservation must be more sharply reasoned than it currently is.

Moreover, these programs wield "soft power" in the world, to use the phrase coined by Joseph Nye.[46] In other words, soft power "co-opts" rather than "coerces," the way hard, or military, power does. Thus UNESCO can "attract and co-opt" as a means of persuasion; soft power has moral not legal or military force. (UNESCO was established "to create the conditions for dialogue among civilizations, cultures, and peoples, based upon respect for commonly shared values."[47]) However, soft power depends on shared values and the duty or obligation to contribute to those values. As we have seen in this chapter, not all values related to cultural heritage are shared universally. And, as I have noted before, there really is no way to enforce the principles of the conventions. Co-opting and even coercion are powerless against any power bent on destroying another culture.

Many countries are nonsignatories to UN declarations. For example, the 1954 Hague Convention, discussed earlier, which aims to protect cultural property during war, has only 128 signatories of the 193 UN member states. The United States did not ratify this convention until 2009, and the United Kingdom has signed but not ratified it, making it "the only major power that has failed to do so."[48] As explained on the website of the UK Department for Culture, Media and Sport, with the passage of The Hague Second Protocol, the British government plans to ratify the convention.[49]

Syria and Iraq ratified the convention in 1958 and 1967 respectively—though not the tougher 1999 protocol, which takes into account "the experience gained from recent conflicts and the development of international humanitarian and cultural property protection law since 1954."[50]

The Second Protocol further elaborates the provisions of the 1954 Hague Convention relating to safeguarding of and respect for cultural property and the conduct of hostilities, and thereby providing greater protection than before. It creates a new category of enhanced protection for cultural heritage that is particularly important for humankind, enjoys proper legal protection at the national level, and is not used for military purposes. It also specifies the sanctions to be imposed for serious violations with respect to cultural property and defines the conditions in which individual criminal responsibility shall apply. Finally, it establishes a twelve-member Intergovernmental Committee to oversee the implementation of the Second Protocol and de facto the convention.[51]

Does the 1954 protocol apply to Islamic State (IS) since they are in Iraq and Syria, two signatories of the 1954 Hague Convention? They have set up their own government, a caliphate. Could key IS members face prosecution by the International Criminal Court? (It is an unlikely prospect.) In the meantime, IS has destroyed hundreds of cultural heritage sites and untold numbers of cultural heritage objects in Iraq and Syria as part of their campaign of genocide and cultural genocide. IS is a perfect example of how all the protective conventions in the world are useless in the face of a "regime" that has its own agenda, and that sees itself as apart from the rules, laws, conventions, and treaties of states.

How are we to preserve cultural heritage under such circumstances? Can we use new strategies that can enhance the work of organizations such as UNESCO? Such strategies include community-based preservation efforts, social media, and the work of *netizens*. In countries like Syria, where it is difficult for the mainstream media to cover the news, netizens, or citizens with Internet and social media access, can provide information by recording events on their cell phones and uploading images to YouTube, for example. The challenge is that it is not always possible to independently verify such information. Another challenge is that such "social" efforts may have no effect against a regime that operates outside all international laws and conventions and is a law unto itself.

There are also international groups like Heritage for Peace, which is a nonprofit organization whose mission is

to support all Syrians in their efforts to protect and safeguard Syria's cultural heritage during the armed conflict. As an international group of heritage workers we believe that cultural heritage, and the protection thereof, can be used as a common ground for dialogue and therefore as a tool to enhance peace. We call on all Syrians of any religion or ethnicity to enter into a dialogue and work together to safeguard their mutual heritage.[52]

In a 2012 paper on the importance of preserving embattled states' cultural heritage, Irina Bokova, director-general of UNESCO, noted:

I am keenly aware that in the context of a tragic humanitarian crisis, the state of Syria's cultural heritage may seem secondary. However, I am convinced that each dimension of this crisis must be addressed on its own terms and in its own right.

There is no choice between protecting human lives and safeguarding the dignity of a people through its culture. Both must be protected, as the one and same thing [*sic*]—there is no culture without people and no society without culture.[53]

Or as Raphael Lemkin put it seventy years ago: "all our cultural heritage is a product of the contribution of all nations."[54]

There are echoes of Raphael Lemkin in Bokova's words. Monumental preservation efforts will require new strategies as well as traditional ones. This chapter has shown the limits of the soft power of international organizations. Nonetheless, without such international organizations and infrastructures, cultural genocide would be more rampant than it currently is. Recent decisions in the International Criminal Court that address the cultural aspects of genocide are encouraging. The ongoing challenge is to develop numerous strategies to protect our heritage. Some new approaches are described in the next chapter.

4 Documenting Cultural Heritage in Syria

PALMYRA
the cradle of ancient civilizations
where monuments inspired by
Greco-Romans and Persians
hold up the sky, and time

stands still, when my hands can't
reach out or encircle the children
who were unable to flee
or to rebuild the walls of bombed out

houses, are unable to light
candles of hope when night and day
are reversed, and a woman who was a wife
and mother lies on the cobbled street

her blood leaving its marks,
while the blind-hearted man
who destroyed so many names
and faces turns away with his rifle

cocked, believes that he is cleansing
Syria in a holy war, cloaked
in ideology, exchanging
a slogan for his soul.
—Marguerite Bouvard, 2015[1]

This chapter explores cultural heritage preservation in Syria, which is embroiled in a civil war that began in 2011.[2] I have selected Syria for two reasons: it has a rich cultural heritage that is in danger of disappearing due to the current strife—much has disappeared or has been destroyed already—and the country's ongoing conflicts make traditional preservation strategies impossible. The country is not safe enough to make it possible for conservators, archeologists, curators, and other professionals from elsewhere to lend in-person assistance to Syrians. Nor can a country where *human* survival is now so precarious expend many resources on cultural heritage preservation. (Yet new strategies are being developed that might prove invaluable in other conflict areas.) To add to an already complex situation, the jihadist group Islamic State of Iraq and the Levant (ISIL)—also referred to as IS (Islamic State) and ISIS (Islamic State of Iraq and Syria)[3]—has been pillaging the country to destroy pre-Islamic buildings for symbolic reasons while stealing antiquities and selling them on the black market to fund further terrorist activities. In August 2015 ISIS beheaded Syrian archeologist Khaled al-Asaad for refusing to reveal the location of ancient treasures in Palmyra.[4] Preserving Syria's heritage in situ is dangerous. The extent to which this war has seeped into the marrow of artistic expression is shown in figure 4.1.

While Syria is not the only country in which civil war is raging, its deep and rich culture and its complex history set it apart from others. Furthermore, the complexities of its situation suggest that there will be no peace there anytime soon. Syria provides a worst-case scenario for safeguarding cultural heritage. This chapter will outline Syria's history and present-day conflicts, illustrate some current approaches to preservation, and suggest new strategies.

Syria abounds in diverse and ancient cultures; two of its cities, Damascus and Aleppo, are among the oldest trading posts and continuously inhabited cities in the world.[5] Syria was part of the Fertile Crescent, also referred to as the Cradle of Civilization, the birthplace of agricultural communities. Wheat was first cultivated there; crop development led to the creation of early trading posts. The Phoenician alphabet, one of the many hallmarks of this region, was the first Western alphabet and a sign of the advanced nature of civilization there.[6] Later, the Silk Road passed through several cities, including Damascus, Homs, Palmyra, and Aleppo—which featured the fourteenth-century Al-Madina Souq, a UNESCO World Heritage Site, much of which has been ruined in the current war.[7] And like some of Syria's immediate neighbors that are also part of the Fertile Crescent—Iraq, Lebanon, Jordan, and Israel—its land is punctuated with the remnants of various cultures, religions, and rulers: Assyrian, Egyptian, Roman, Jewish, Christian, Muslim, Umayyad, Abbasid, Mongol, Ottoman, and European. Robin Yassin-Kassab and Leila Al-Shami say that Syria is "pocked with tells, hills made over millennia of human habitation—the pebbles beneath your feet are not pebbles but the shards of ten million pots manufactured and discarded generation after generation."[8]

Tragically, "shards" describes Syria's current situation all too well as a country in pieces—its infrastructure, its very physical presence, is in rubble, figuratively and literally, because of the shelling it has endured from all the warring parties there.

Following the fall of the Ottoman Empire (after 1914) and through World War I and immediately after it, the French and British divided most of the Fertile Crescent between themselves (with the assent of soon-to-be overthrown imperial Russia and the agreement of other European countries). First they authored the (initially) secret Sykes-Picot Agreement of 1916 (see figure 4.2);[9] it later became the French Mandate for Syria and the Lebanon, which the League of Nations passed in 1923. Britain claimed most of Mesopotamia (modern Iraq), southern Syria, and Transjordan, while France was given control of the rest of Syria, Lebanon, and portions of southeastern Turkey. Palestine was placed under Britain. The division was strategic: Britain wanted to control the land route to India and to have a cheap and steady supply of oil for its naval fleet, while France sought to preserve its ties to Syrian Catholics, secure an economic base in the eastern Mediterranean, and prevent Arab unrest from spreading to their French North African empire.[10] The Arabs were left out of the discussions, and in fact did not originally know about Sykes-Picot. Significantly, and tellingly, the French and British motivations seem to have had little or nothing to do with the Arabs—their political, social, economic, or aesthetic interests—and everything to do with their own interests, mostly economic. Any long-term thinking would have seen that to suppress a culture the way they did, coming from outside powers, would, in the long run, seriously irritate—and increasingly infuriate—the indigenous culture. The French and British chose what eventually proved to be short-term profit for long-term trouble. Instability was destined to ensue.

In fact, attempts by Amir Faysal Ibn Husayn and others in April 1920, at the San Remo Conference, to establish a fully independent Arab state were unsuccessful. The French promised Syria a certain amount of autonomy, but in exchange, Syria had to recognize the independence of Lebanon. The Party of Arab Independence, supporters of Faysal, did not agree with the terms; in their view Lebanon and Palestine needed to remain part of Syria. The party further rejected a Jewish national home in Palestine, a plan put forth in the

Figure 4.1

a) Zaher Omareen, *Faceless 82*, n.d. b) Sulafa Hijazi, *Untitled*, 2012. c) Khalil Younes, *Untitled 5*, 2011. Permission of the artists. These images originally appeared in *Syria Speaks: Art and Culture from the Frontline*, ed. Malu Halasa, Zaher Omareen, and Nawara Mahfoud (London: Saqi Books, 2014).

Figure 4.1 (continued)

Balfour Declaration of 1917. The San Remo Conference was where the terms of the French Mandate were worked out; it "gave" Palestine and Mosul to the British. Britain, in turn, agreed to evacuate its army from Syria.[11]

It is noteworthy that the proposals—who got what land; who answered to whom; who ruled and who was to be subservient—were made by powers that had their own interests foremost in the planning. The kind of narrow self-interest exhibited in these dealings was bound to lead to trouble. What is occurring in Syria today—and in so much of the Middle East—is clearly a direct result of ignoring history.

Under Ottoman rule Syria had been made up of a number of subdistricts or administrative units—known as *wilayas* in Arabic and *vilayets* in Turkish. For example, Aleppo, Homs, and Damascus were each vilayets. The districts, however, contained more territory than do their modern namesakes. For example, the Damascus region included much of modern-day Lebanon. The Syrian borders were the Taurus Mountains on the north, Aqaba and Sinai on the south, the Syrian desert on the east, and the Mediterranean Sea on the west. Today that land encompasses Jordan, Israel/Palestine, Lebanon, and Syria. The Sykes-Picot Agreement drew lines through the vilayets and thus severed old community ties. Critical to an understanding of today's conflicts, new national identities and social classes came to "challenge[,]

Figure 4.2
The Sykes-Picot Line Map (1918). Permission of the National Archives (UK)

and even replace, the identities of clan, tribe and religion."[12] Syria was subdivided into five parts: Lebanon, Syria, Jabal al-Druze, the Sanjak of Latakia, and the Sanjak of Alexandretta. Taking discrete and relatively homogenous areas and fairly arbitrarily splitting them up by land mass with no sensitivity to culture was another blueprint for trouble.

Sykes-Picot and, later, the Treaty of Sèvres (August 1920)—which abolished the Ottoman Empire and obliged Turkey to renounce all rights over Arab Asia and North Africa—affected the Kurds. The divisions set forth in Sykes-Picot caused Kurdistan to be divided. However, during the French Mandate, Kurdish national identity was recognized. In fact, the French wanted to prevent an upsurge of Arab nationalism and so supported minority groups. The Treaty of Sèvres granted Kurds and Kurdistan the right to independence—but it was not meant to be; the Turkish war of independence caused the treaty to collapse.[13] When Syria gained independence from the French, the Kurds lost their minority rights. Today Kurdistan spans four countries: northern Syria, southwest Turkey, northern Iraq, and northwestern Iran. The Kurds have struggled for autonomy for decades in each of these countries. The Syrian government does not consider them to be a national or ethnic minority. With some thirty-two million Kurds worldwide, their identity has remained strong, even against pressure for them to assimilate in Syria and elsewhere. They are the largest ethnic group without a state.[14]

The Kurds are not the only minority group that has suffered, and their situation is but one of many destabilizing situations in the Middle East. To preserve cultures and the artifacts of cultures, the region needs to preserve its peoples. But with so many conflicting interests, military conflict has become inevitable, especially given all of the religions and cultures in the region.

The French lost little time in taking over Syria; they deposed Faysal in July 1920. (The British made him King of Iraq the next year.) The French then divided Syria into segments, as mentioned earlier. Lebanon was created as a separate state, becoming its own country for the first time. The plan was favorable to the Maronite, Greek Orthodox, and Catholic Christians, but most—though not all—Muslim groups there preferred to be part of Syria. The French hoped that with these splits, Lebanon would remain a Christian-dominated French power in the Middle East with no desire to become an independent nation. The French similarly followed a divide-and-conquer approach in Syria, creating many separate states. Each state had a local governor supported by a French advisor. The French did little to help Syria transition into an independent country. In fact, France imposed constitutions on Syria and Lebanon that stayed in place until the British forced France to leave Syria in 1946.[15] "The outcome was that Syria emerged...as a unitary state with very little experience of unity....[The French legacy] was almost a guarantee of Syria's political instability."[16]

In 1945 the Arab League or League of Arab States was created in Cairo with the collaboration of the Kingdom of Egypt, the Kingdom of Iraq, Lebanon, the Syrian Republic, Saudi Arabia, Trans-Jordan, and North Yemen as member states, in a show of Arab nationalism. (The Palestinian Liberation Organization [PLO] became a member in 1976; today most Arab countries are members.) In 1948, Syria joined its fellow Arab League members in attacking Israel (unsuccessfully). Other attacks against Israel by neighboring Middle East countries would follow in 1956, 1967, and 1974. Israeli-Syrian relations form just one layer of many internal and external Syrian conflicts.

M. E. McMillan points out that Syria's quest—after it had gained independence from France—was to reclaim its former lands. This has prompted the Syrian government "to intervene in Lebanon's civil war (1975–1990), to ally with the Shi'a group Hizbullah, and to stay on as an occupation force until they were forced out...following the assassination of former Lebanese prime minister, Rafiq Hariri, on Valentine's Day 2005."[17]

Syria has had continual internal strife since the end of the French Mandate. In 1949, three military coups initially failed to give Syria stable leadership. The coups led to the rise of the socialist Ba'ath Party—and to the rise of the army that has continued to be a central part of the Syrian government. By 1957, a well-organized Syrian Communist Party seemed poised to take over the country. Fearful of a takeover, Syrian President Shukri al-Quwatli and Prime Minister Khaled al-Azem met with Egyptian president Gamal Abdel Nasser, seeking an alliance with Egypt. As a result of those talks, in 1958 Syria merged with Egypt to become the United Arab Republic. However, it was an unpopular move in Syria because Nasser ultimately wished to absorb Syria, not to merge with it. Thus in 1961 the Syrian army staged a coup d'état that ended the union with Egypt and restored the Syrian republic.

A coup d'état in 1963, also known as the 8 March Revolution, put the Ba'ath Party in power. The coup was staged by the military committee of the Syrian Regional Branch of the Ba'ath Party. Michel Aflaq, one of the founders of the Ba'ath Party, supported the coup. From then until now, Syria has been a military-run state whose leaders have exercised repressive power.

In 1966, radical Ba'athists overthrew the existing government. This shift ultimately allowed Hafez al-Assad, who had been minister of defense and commander of the armed forces, to seize power in November 1970. He was elected president in March 1971, and remained so until his death in 2000. His election further cemented military rule in Syria. Assad was a member of the Alawites, a sect of Shi'a, who are indigenous to Lebanon and Syria. He soon clashed with the Sunni Muslim Brotherhood.

Hafez al-Assad was an authoritarian ruler who brutally repressed all opposition. At the same time, he sought to regain Syria's former territories and establish stability in the region. While these goals may seem antithetical, they were not for Assad. He did not want Lebanon's civil war to spill over into Syria, and he knew that some Lebanese would welcome his interference. In 1976, when Assad intervened in the Lebanon civil war, he had many reasons for doing so:

- the land had originally belonged to Syria
- Israel and its Lebanese Christian allies wanted to establish a Christian ministate in southern Lebanon that would act as a buffer to Muslim extremists
- he sought any opportunity to aggravate Israel (which had taken the Golan Heights from Syria in the Six-Day War in 1967)
- Iraq, Syria's enemy, had joined the Lebanon unrest by supporting Christian and other militia
- The presence of European and Israeli forces in Lebanon threatened Syria's sense of security.[18]

However, Assad's ultimate goal was self-preservation, which he was willing to achieve by any means. Yassin-Kassab and Al-Shami point out that the Syrian army first prevented a Palestinian victory in Lebanon and then slaughtered Palestinians in the Lebanese camps.[19]

In 1979, Assad recognized Iran's new government after the Iranian revolution, and the next year he supported Iran in the Iran-Iraq War. (That early support has borne fruit; in 2013, Iran provided Bashaar al-Assad a $3.6 billion credit line, separate from military aid.[20]) The following year, in 1980, an assassination attempt against Hafez al-Assad resulted in brutal repression of those who opposed him. Two years later in 1982, there was another Syrian revolt, by the Muslim Brotherhood. Assad managed again to quash the uprising and the government killed many thousands in revenge. None of Assad's efforts were instrumental in furthering peace in Syria. While he could repress rebellion, his brutal and murderous repression ultimately engendered instability.

In the meantime, Hezbullah (The Party of God) was emerging in Lebanon in the 1980s as a political force. The party pledged allegiance to Ayatollah Ruhollah Khomeini. Hezbullah used suicide bombs as a weapon of choice against Western targets, such as the suicide truck bomb that killed 241 U.S. Marines and 60 French soldiers in 1983. Hafez Assad began a complex relationship with Hezbullah that has continued under his son Bashaar.

The senior Assad's desire to take power in Lebanon remained. In 1990 he deposed General Michel Aoun and extended Syria's military authority in Lebanon. On June 10, 2000, Hafez al-Assad died, after thirty years in power. Parliament revised Article 83 of the constitution to reduce the minimum age of the presidency from 40 to 34—the age of Hafez's son Bashaar. On July 17, 2000, Bashaar was elected president. A brief period of "reform" was followed by a newly repressive government. In 2011, following on the heels of the Arab Spring movement,[21] a revolution began in the streets among Syrians from all backgrounds. It was inspired by the social and economic disparities that had existed for decades under the two Assads. Nevertheless, Bashaar is still in power and he continues to quash the country's many opposition groups— and his efforts have the generous support of the Russians and Iranians.

What is now a civil war began as a series of protests. On February 17, 2011, 1,500 people demonstrated in Damascus after traffic police beat up the son of a local merchant. The Interior Minister went to the scene and asked whether the participants were holding a demonstration; when they said no, the crowd was left alone.[22] However, a second Day of Rage was organized for March 15. Social-economic inequality and widespread political oppression in Syria inspired people to protest Assad's repressive regime. Thousands turned out for the second demonstration. Protests continued around the country and it wasn't long before clashes ensued. In April protests spread to some twenty cities. By the end of April, the Syrian army had begun military attacks on the towns. By the end of May, thousands of people had been arrested, and many were killed. Nonetheless, the protests continued to grow; in July large ones took place in Hama, where in 1982 the Muslim Brotherhood had organized an armed uprising.[23] However, Hafez al-Assad was able to repress the Brotherhood.

Protests escalated to armed conflict, and Bashaar used violent means— including cluster bombs and chemical weapons—to repress those who were against his regime. Over the past five years, the war has been fought among a variety of factions: the Syrian government, Free Syrian Army fighters, Salafi jihadist groups, the Kurds, ISIS, and many others. In 2015 the Kurdish YPG (the People's Protection Units) joined forces with Arab, Assyrian, Armenian, and Turkmen groups to form the Syrian Democratic Forces.[24]

An overview of the various religious and ethnic groups in Syria provides further context for the country's ongoing struggles. In the mid-1990s, Nikolaos van Dam reported that over 80 percent of the country was Arabic-speaking and nearly 70 percent of the Arabic population was Sunni Muslim—Sunnis were nearly 58 percent of the whole population. The major *religious* minorities were Alawis (including the Assad family), Druzes, Isma'ilis, and Greek Orthodox Christians—who constituted the most significant Christian community.[25] The primary *ethnic* minorities were the Kurds, Armenians, Turcomans, and Circassians. Further distinctions can be made: the Kurds, Turcomans, and Circassians were predominantly Sunni; the Armenians were mostly Christians, and so on. Thus some groups represent ethnic and religious minorities, some only ethnic, some only religious. Religious minorities can be further divided into Arabic- and non-Arabic-speaking groups.[26]

Population figures are not easy to come by today. After five years of civil war, at least 250,000 Syrians have died and millions have fled from the country. The United Nations estimates that the population was 24.5 million before 2011, and that it is now 17.9 million.[27] A 2013 article by Max Fisher in the *Washington Post* reproduces a map of Syria created by the Columbia University Middle East Institute Gulf/2000 Project[28] which creates ethnographic and cultural maps of several Middle Eastern countries. It includes many groups not included in van Dam's list including Copts and Assyrians. There are separate enclaves as well as mixed communities.

Fisher points out that the maps have strategic implications for where Bashaar al-Assad is strongest (the Alawite regions) and weakest (the Kurdish regions). Thus, one can trace the killings that are along sectarian lines. He further suggests that the Syrian war

began for political reasons—people protesting dictatorship, the dictatorship overreaching in suppressing those protests by force, things spiraling out of control until it's civil war—but that the fighting is causing people to retreat to sectarian identities and antagonisms, to make the old divisions deeper and more vicious. Sectarian conflict, after all, can have its own self-reinforcing logic: Alawites are bonding together in part because they fear, not without reason, that they'll be slaughtered in Sunni revenge killings if Assad loses. Sunnis see Alawite militias forming and thus perceive all Alawites as their enemies, so they start attacking members of that religious sect, which makes other Alawites more likely to form in-group militias. And on. [*sic*][29]

The conflicts are even more complicated than that because there are outside groups involved in the war as well. Bashaar al-Assad has received support from Hezbullah, the Lebanese Shiite militant movement that started in the early 1980s in the midst of the civil war in Lebanon. Meanwhile, Hamas, the Sunni Palestinian group, has trained and armed Syrian rebels. The two groups have long been allies against Israel, so they have "agreed to disagree" about Syria.[30] Add to these the terrorist groups that have risen in recent years, such as al-Qaeda and ISIS, and you have a lethal mix of groups all with their own motivations for being in Syria.

The war has now gone on for more than six years. What do we know about those civilians who could not escape from Syria? The Violations Documentation Center is a network of activists inside Syria that records the names of victims and the cause of their deaths. Outside Syria, the United Nations estimates that 13.5 million people need humanitarian assistance. And Physicians for Human Rights has "documented 336 attacks on at least 240 medical facilities across the country. The attacks resulted in the deaths of 697 medical personnel" between March 2011 and November 2015.[31] The situation is grim.

As of this writing, Syrian forces have attacked the Kurds in al-Hasakah (northeast Syria) who have been trying to retake territory—the Kurds' semi-autonomous region—from ISIS, as Bashaar al-Assad tries to gain control of all of Syria. Meanwhile, the Russians and Iranians are coordinating their efforts to support Assad. Russia (and formerly the Soviet Union) and Iran have deep and long-standing ties to Syria. The civil war in Syria is becoming a proxy war with a number of international players.

A brief summary of Syria's history cannot do justice to its complexities and nuances. However, it does suggest that the political situation will not be resolved soon. Ayse Tekdal Fildis, M. E. McMillan, and other scholars rightly suggest that the problems now plaguing Syria will be of long duration. The Sykes-Picot Agreement is just one of many ill-considered plans made by outsiders who wanted to control parts of the Middle East for their own gains. Significantly, Britain and France failed to facilitate the creation of a post-imperial government that might have prevented the ethnic, religious, and tribal divisions that plague Syria today. Unfortunately, history may repeat itself with the involvement of foreign powers that are again trying to control Syria's future.

What is to be the fate of the Syrians? and of their cultural heritage? Assad's forces as well as ISIS have been practicing genocide; what about the cultural genocide that is also taking place?

* * *

In chapter 3 we considered the concept of genocide, and the person who coined that term, Raphael Lemkin. To Lemkin, the term denoted the large-scale massacre of people, and also the destruction of their culture. Clearly, his conception of genocide encompassed cultural genocide as well. For reasons of political expediency, the Genocide Convention of the United Nations defined the term more narrowly than did Lemkin. But he continued to write about cultural genocide for the rest of his life.

Lemkin spent decades working on *Introduction to the Study of Genocide*, a book of which only thirteen of the proposed sixty-three chapters were completed. The fragments were published only a few years ago—some fifty years after his death.[32] He undertook his research at the same time that he was meticulously documenting the imposition of Nazi occupation laws throughout Europe during World War II, which resulted in his exhaustive study, *Axis Rule in Occupied Europe*.[33] The juxtaposition is revealing: Lemkin was documenting the Nazi occupation while simultaneously trying to place Nazi genocide in a broad historical, legal, economic, and cultural context. He was also trying to further clarify his concept of genocide (see chapter 3). By documenting the long history of genocide from an interdisciplinary perspective, Lemkin hoped that people would understand its components—national, racial, religious, and ethnic—which derived from his premise that "genocide is an organic concept of multiple influences and consequences."[34] For Lemkin, research and documentation made it possible to remember—and learn from—the injustices of the past.

Documentation was a key component of his work. When Lemkin moved to the United States in 1941, he was surprised that "denial was still the prevailing sentiment in the United States" for Hitler's crimes.[35] Americans were slow to believe the extent of the Nazi occupation— Lemkin needed to prove it to them. He did so successfully with *Axis Rule*. However, he never completed the *Introduction to the Study of Genocide*, which he envisioned as a companion work.[36] Nor is it likely that he could have done so; the scope of the project was probably too large, Lemkin lacked the time and money to work on the book full time, and he died at age fifty-nine.

Lemkin's dogged and thorough approach to uncovering injustice is relevant today. Documentation is still a critical tool in exposing human rights and cultural heritage abuses, but it is now easier to document in "real time," and with new technologies. What is the role of documentation in the ongoing civil war in Syria?

*   *   *

[W]hen a nation or a group is destroyed it is prevented from making cultural contributions.
—Raphael Lemkin[37]

Palmyra was built with stone and mortar. It will be rebuilt with computers and drones.
—Karen Leigh[38]

The documentation of heritage dates back to ancient times. Thanks to the scribes in Assyria—who recorded on clay tablets voluminous information about their culture—we are as knowledgeable as we are about the ancient Near East. Fast-forward nearly four thousand years to the same region of the world: documentation continues to be important for the safeguarding of culture. Now there are many ways to document information, though not all will be as durable as clay tablets have been. But what we might lose in the long run—permanent records—we have in the short run, thanks to our many modes of communication such as smartphones and email. The inherent paradox of our ubiquitous digital technologies is that a great deal of what they capture will disappear, a topic covered in chapter 7. For now, we will focus on how the current panoply of technologies, combined with social networking, can capture Syrian heritage.

One way to try to preserve cultural heritage that is endangered—or that has already been damaged—is by marking heritage sites. The International Committee of the Blue Shield (ICBS) grew out of the 1954 Hague Convention (see chapter 3). One of the recommendations of the convention was to mark heritage sites with a depiction of a blue shield that would designate protected objects and buildings (see figure 4.3). The ICBS was founded in 1996

to promote the protection of cultural property (as defined in the Hague Convention) against threats of all kinds and to intervene strategically with decision-makers and relevant international organisations to prevent and to respond to natural and man-made disasters. ... ICBS works for the protection of the world cultural heritage by coordinating preparations to meet and respond to emergency situations as well as post-crisis support. And it promotes good standards through risk management

Figure 4.3
The Blue Shield. Courtesy of the International Committee of the Blue Shield

training and awareness-raising campaigns for professionals and the general public. Its unrivalled body of expertise allows the organisation to collect and share information on threats to cultural property worldwide, thus helping international players to take the appropriate measures in case of armed conflict or disaster.

ICBS intervenes as an advisor and cooperates with other bodies including UNESCO, ICCROM and the International Committee of the Red Cross (ICRC). Finally in emergency situations, ICBS encourages the safeguarding and the restoration of cultural property, the protection of threatened goods, and helps the professionals from the affected countries to recover from disasters.[39]

Thus, marking sites is one form of documentation. But documentation also includes other activities such making lists to compile inventories. The information must be accurate, accessible, and secure to whoever is responsible for securing the endangered heritage. In his "War and Heritage: Using Inventories to Protect Cultural Heritage," Peter Stone describes and analyzes the complex challenges he faced in Croatia in the 1990s, and in Syria today.

We also need a debate about how widely these lists should circulate. In a conversation I had in 1999 with the minister of culture for Croatia, he noted that on the eve of war in the former Yugoslavia, Croatia had, as required by the 1954 Hague Convention, produced a list of property to be protected and sent it to UNESCO. He told me that in the fighting that followed, every site on the list was targeted by opposition forces. While debate about protecting cultural property during conflict mostly relates to unintended damage, in the Balkans conflict cultural property was targeted as part of a political strategy. We will never know (but can probably guess) whether the sites would have been targeted if the list had not been produced. Identifying sites on a list at least provides evidence to the world judiciary for the trials of those responsible for intentional damage—as indeed happened in the prosecutions for the targeting of the World Heritage Site of Dubrovnik. ...

A third issue is when and by whom lists are produced. The publication of a definitive list for Syria was delayed, while fighting continued, as four different lists—produced by four different groups with differing levels of contact with Syrian experts—were compared and collated. The different lists had different English spellings and therefore different records of the same sites. Some lists had good GPS data; others less good. Some had explanations of the importance of sites; others did not.[40]

Despite these obstacles, the consensus is that a system of creating lists of heritage sites and marking them is in most cases more beneficial than harmful.

The ICBS has representatives from five organizations: the International Council on Archives, International Council of Museums, International Council on Monuments and Sites (ICOMOS), International Federation of Library Associations and Institutions, and Coordinating Council of Audiovisual Archives Associations.

Two declarations have been issued over the past twenty years to further amplify the role that heritage professionals can play in response to natural and human-made disasters: the Radenci Declaration on the Protection of Cultural Heritage in Emergencies and Exceptional Situations (arising from meetings held November 12–16, 1998); and the Seoul Declaration on the Protection of Cultural Heritage in Emergency Situations (meetings held December 8–10, 2011). The Radenci seminar was held in Radenci, Slovenia, to train personnel to provide aid following armed conflict or natural disasters. One of the case studies presented was on war damage in the former Yugoslavia[41] (in 1999 ICBS organized missions to Kosovo in association with UNESCO and the Council of Europe).[42]

The Radenci and Seoul Declarations came about because threats to the preservation of cultural heritage have been "exacerbated by an increasing frequency and intensity of disasters, and its full range of collateral effects."[43] The war in Syria continues to have collateral effects. ICBS promotes preservation, but in addition to training volunteers in disaster response and sponsoring conferences, what other steps can be taken to protect world heritage?

UNESCO launched The Emergency Safeguarding of the Syrian Cultural Heritage Project on March 1, 2014, a three-year endeavor. It is supported by the European Union with the additional assistance of the Flemish government and Austria. The project, based in Beirut, has three aims: to monitor and assess Syrian cultural heritage through continuous documentation, to mitigate the destruction and loss of Syrian cultural heritage, and to sponsor international communication and awareness-raising efforts through technical assistance.[44]

The first of these—monitoring and assessing cultural heritage—may be doable, though doing so places the monitors in danger. How can anyone be safe in a country with constant bombardment and so much instability? The second aim—mitigating destruction and loss—is even more problematic, for the same reason. The recurring idea that conventions and laws mean nothing to aggressors, who will do what they like—bomb whatever they wish to destroy and steal whatever they want—to achieve their aims, makes it practically impossible to mitigate loss. And the third aim, to sponsor international awareness-raising, may also be doable, but to what purpose? Being aware of the problem does not solve it, nor does it stop the destruction and loss of cultural heritage. As noted a number of times in this book, understanding the problem of the loss of cultural heritage is not difficult; being able to do something about it may be impossible, especially in war zones in which combatants play by their own rules.

One approach to preserving heritage is to photo-document it. That way, even if the original disappears, or is destroyed, there will be a record (which itself must be preserved). The Hill Museum & Manuscript Library (HMML) has been microfilming and digitizing medieval manuscript collections in monasteries since 1965. They began their work in Europe and later moved on to North Africa and the Middle East. They catalog the images and archive the digital data. HMML began digitizing manuscripts in Syria in 2004 and continued to film there until 2012.[45] Such partnerships can assure that materials, archives, and objects are documented before they are damaged or destroyed, but these projects are expensive—and not all of them are sustainable. HMML receives support from a number of foundations. Also, this solution presumes a stable situation in which the preservation actions can be carried out. But what about trying to preserve works of art or architecture in volatile zones? It is risky at best and downright deadly at worst. The technology is there for such preservation; it only needs conditions that allow us to use it.

CyArk is a nonprofit organization in California that digitally captures world cultural heritage sites "through collecting, archiving and providing open access to data created by laser scanning, digital modeling, and other state-of-the-art technologies" (www.cyark.org). It was founded by Ben Kacyra, an Iraqi-American civil engineer, as a response to increasing threats to heritage sites all over the world. CyArk collaborates with a number of cultural heritage organizations. Recently CyArk has partnered with Yale University's Institute for the Preservation of Cultural Heritage and the International Council on Monuments and Sites on Project Anqa, an emergency program for recording high-risk heritage in the Middle East and North Africa. It deploys "teams of international professionals, paired with local experts to digitally document at-risk sites in 3D before they are destroyed or altered."[46]

Spatial archaeometry studies the properties of archeological materials. Geospatial and satellite imagery can be used to observe an archeological site from a great distance; drones can be used to record information about a site at close range.[47] These technologies can be used to assist governments in the protection of their heritage by mapping looting patterns, or providing proof of pillaging. Such technologies can reduce the illicit trade in antiquities, and provide useful data for security experts. The problem, however, is that mapping looting patterns and giving proof of pillaging is *post hoc*: the looting and pillaging are already done. Documenting it does not nearly guarantee that what has been looted will ever be returned.

Additionally, the whole notion of halting looting is complicated by the widespread locations of vulnerable sites, countries' reluctance to fund anti-looting programs, their inability to identify looted materials and to stop the looters, and other problems. All of the conventions in the world do not constitute international, enforceable laws. They are merely guidelines. To stop looting, countries need to remove the incentive to loot: profit. As long as profit can be had from the theft of heritage objects, looting will continue. That is why collaborative and proactive activities hold the most promise.

Another motive, however, is at play—an ideological one: religion. In the name of one's deity, a person (we can call him "a fanatic") can justify the destruction of *anything*. Removing this motive is futile, for it means asking people to deny their more deep-seated beliefs—to deny their faith. Preventing the destruction of cultural heritage may be hopeless. So it is especially important that we document the signs of cultural heritage that we wish to preserve. Again, collaborative initiatives are critical.

Other Initiatives

The American Schools of Oriental Research (ASOR), which was founded in 1900, promotes the study of the Near East through a number of programs and partnerships. In August 2014, ASOR began its Cultural Heritage Initiatives program to document, monitor, and report on cultural heritage damage in Syria. This program is an international collaboration with participants from Syria, Iraq, the United States, Canada, England, France, Germany, Lebanon, and Jordan. It has developed seven strategies:

• increase public awareness about the destruction of sites in Syria
• facilitate in-country documentation to "identify and document destruction as it occurs in real time"
• use satellite remote sensing to monitor sites, analyze past images for changes over time, and verify and monitor current damage at cultural heritage sites
• create a cultural heritage inventory to build a map and inventory of items that includes archeological sites, museums, libraries, archives, and other historic sites
• develop emergency response networks, working with locals to teach them how to safeguard heritage
• work on long-term preservation projects that will draw on the current initiatives and can take place once the situation in Syria is stable, and
• maintain a cultural heritage bibliographical database that includes sources on cultural heritage sites in Syria in Arabic, English, German, French, and Italian, and other languages in which sources are found.[48]

One small organization is worth mentioning: Yazda: A Global Yazidi Organization. The Yazidis, a Kurdish-speaking ethno-religious minority in northern Iraq and Turkey, were recently the victims of ethnic cleansing by ISIS. They were forced out of their homeland in the Sinjar District of the Nineveh Governorate, Iraq, and have subsequently fled to parts of Kurdistan (including Syria) as well as to Europe. Yazda has a genocide documentation project that will become a historical record of the attempted extermination of Yazidis. It is important to document such history as close to its occurrence as possible, if it is to be captured and preserved.[49]

A war as complex as the one in Syria calls for an array of preservation strategies. Preservation in a peaceful place is challenging enough. The many difficulties discussed in this chapter would seem to make preservation initiatives impossible. Given that personal safety and health must be of preeminent importance to Syrians, home-grown preservation initiatives are truly miraculous. That people are willing to put their lives on the line to preserve elements of their culture demonstrates the importance of heritage in our lives. Preservation is not easy; it requires funding and coordination. For how long will external funding be available for Syrian preservation initiatives?

This chapter has described just a few of the many initiatives currently under way. It remains to be seen how much of Syria's cultural heritage will be preserved. Given current conditions there, such preservation is a truly monumental endeavor.

III Approaches to Preservation

5 Collecting as Preservation

[C]ollecting is the accumulation of tangible things.
—G. Thomas Tanselle, from "A Rationale of Collecting" (1998)[1]

This chapter explores the nature of private and institutional collecting and its relationship to preservation. The intersection of preservation and collecting will be considered from the vantage point of librarian and art collector Jean Brown, as well as some of her forerunners who also gathered materials in new areas or had new approaches to collecting. Brown collected Dada and Surrealist art and their offshoots, Fluxus, mail art, and concrete poetry.[2] Her materials pose many preservation challenges as items representing these art movements were ephemeral in composition or conception, or both. Her collection anticipates some of the problems inherent in digital media. Since some of the Fluxus and other forms of art were seen by their own creators as ephemeral, the ethics of collecting it—the necessary prelude to its preservation—raises two challenges: Should we save art that artists are willing to let die? And, how should we preserve it?

But first, what *is* collecting? There are many definitions and thousands of books written about it.[3] I have selected G. Thomas Tanselle's definition of *collecting* to begin this chapter because it is succinct and descriptive. He has intentionally left out the motivations for and end results of collecting. His justification is that no definition could fully "encompass all the causes and all the results of collecting."[4] It is an appropriate definition for this chapter because cause and effect are not primarily of interest here; what is important is the fact that the activity of collecting results in the creation of holdings that find their way into—or are built within—institutions and are subsequently studied, used, and preserved. Collecting is at the core of heritage institutions, and the nature of particular collections differentiates one institution from another. Collections define an academic institution, especially when they support classroom teaching, research, symposia, exhibitions, and other pursuits. Preservation makes collecting possible and collecting makes preservation necessary. Also, collecting itself is a form of preservation.

Tanselle's elegant essay "A Rationale of Collecting" stands out in a crowded field. Of the writing of books and articles on collecting there is no end. This fact was brought home to me when I was doing research at the Getty Research Institute Library. More than five hundred books on the collecting and provenance of books and art line one wall.[5] This wall's range is not all that the library holds: other materials on the subject are in special collections, in the general stacks, and in storage, but it was striking to see so many books on the subject shelved in *one* place. A recent Online Computer Library Center (OCLC)[6] WorldCat subject search on "collecting" retrieved millions of records. Book collecting is the largest category (over seven million hits) followed by art collecting (over six million). Even when the search is refined to winnow copies of the same titles, there remains a large number of items on the subject. There are scholarly journals devoted to the subject (such as *The Book Collector* and the *Journal of the History of Collections*) and many organizations, clubs, and societies—scholarly and hobbyist. But it is not just literature about collecting that interests us. *Antiques Roadshow* and a host of other television shows such as *American Pickers* and *Market Warriors* appeal to treasure hunters—or those just curious about the value of objects in their possession. Those who want to buy and sell objects or build collections can turn to a seemingly infinite online marketplace where eBay, viaLibri, BookFinder, AbeBooks, Amazon, and many other relevant Internet sites reside.

Collecting is about more than buying and selling things. Objects in general, and collections in particular, are part of our identities. Tanselle quotes a section of Lord Eccles's book *On Collecting* in which Eccles describes how during the London blitz in 1940–1941, people whose homes were bombed regretted the loss of their things far more than the loss of their houses. Similarly, a colleague whose home was destroyed in the 1992 Siege of Sarajevo observed that she missed her family photo albums the most. In losing the albums she lost an important part of her family history. Survivors of natural disasters often express similar sentiments when they are interviewed by the media. For most people dwellings are merely protective, and they are acquired "all at once." But possessions and collections are built over many years, or even generations, the result of hundreds of individual decisions, each one of which reflects someone's personality, desires, and values. It is no wonder, then, that collections usually mean more to their owners than do the buildings that house them.

We are deeply connected to our things, as Sherry Turkle illustrates in *Evocative Objects: Things We Think With.*[7] The chapters are essays written by her colleagues about objects of significance to them. The objects include stuffed animals, jewelry, appliances, musical instruments, tools, and comics. The variety is matched by the variety of stories about them. Turkle notes that the meaning of objects "shifts with time, place, and differences among individuals."[8] There is a distinction between cherished objects and collections: all collections are made up of objects, but a group of objects is not necessarily a collection. A collection is an accumulation of "tangible things" with a purpose. It is possible to cherish each item in a collection, but sometimes it is the collection as a whole that is treasured.

Collections can lead to the recapturing of family histories. In *The Hare with Amber Eyes: A Hidden Inheritance*, Edmund de Waal describes his quest to learn about his Russian-German-French-English-Dutch-Jewish ancestors— and the tragedies that they endured before and during World War II. His journey began with a netsuke collection that he inherited from his great-uncle Ignace Ephrussi (Iggie), in Tokyo. The netsuke were purchased in the 1870s in Paris by Iggie's cousin Charles Ephrussi who collected Japanese objects, among many other things. This remarkable collection of small carved items, which were easy to hide, survived wars and many relocations— a poignant remembrance of the many losses and upheavals endured by de Waal's family. The Ephrussi family history is inextricably related to their netsuke collection.

Every collection tells us something about the culture in which it was created—and the collectors. Objects convey layers of information. And while some may judge a collection of early printed books or Old Master paintings to be more "valuable" than a collection of nineteenth-century greeting cards, or outsider art (created by artists with no formal training, outside the established art world), we can learn from the old and the new, and the prestigious and the modest. Every collection, no matter its fiscal value, contains information of many kinds: about the creators of the objects, the times in which the objects were made or collected, the reasons why people create and collect, the institutions that are often the final repositories of objects and collections and the scholars who use them, the things they say or write about them, and so on. Of course, what is valued at any given time must also be measured against prevailing tastes. And tastes change constantly, in books, in art, in fashion, in scholarship, and in most areas of collectibles. The value of an object is sometimes inverse to what might be expected; a quilt by an unknown craftsperson might command a higher price than an etching by a well-known artist. Changes in scholarship have an impact on the value of a collection to a scholar, and much contemporary scholarship depends on collections of ephemeral objects. There are many measures of the value of an object—or a collection.

This last point is worth expatiating on. There is an inherent contradiction in collections of ephemera—objects that by their very nature, from their original conception and manufacture, were not meant to last, let alone be collected. (Though not all ephemera was intended to be transient; for example, baseball cards or paper dolls were ephemeral items that were designed to be kept.[9]) But people by their basic nature want to preserve things that have meaning to them. They may see this meaning as having a purpose to others as well, but that might not be the reason for their collecting. People's reasons for collecting are not germane here; the fact that they collect *is* germane. And the fact that these collections wind up in institutions and become the objects of research is central to this discussion, which is to show that even the humblest materials can become central to scholarship. Individual items may not tell us much, but when they are assembled into coherent groups—collections—their potential for imparting knowledge becomes great. And the contradiction that I spoke of—the perpetuation of large numbers of items originally conceived of having only short lives—has created long-term markers of culture.

Some collectors are visionary and they recognize the significance of objects before others do. One example is Electra Havemeyer Webb (1888–1960). She was born to a wealthy family, and her parents, Henry Osborne (H.O.) and Louisine Havemeyer, were avid collectors of Impressionist and Asian art, as well as many other areas that were popular in the late nineteenth century. (The bulk of their collection was bequeathed to the Metropolitan Museum in 1929; see figure 5.1.) Electra developed her own interest in collecting as a child and was a serious collector by 1911, long before Henry Francis du Pont, Henry Ford, and others were collecting in related areas of American material culture.[10] An early acquisition was a cigar store Indian, and that was the beginning of her life-long fascination with Americana.

Figure 5.1
Louisine Havemeyer and Her Daughter Electra, 1895, by Mary Cassatt. Pastel on wove paper, 24" × 30 1/2". Photography by Bruce Schwarz. Permission of the Shelburne Museum

From her husband's family's extensive properties in Vermont, she was able to acquire land and buildings in which to house her collections.

Electra Havemeyer Webb took pleasure in collecting and preserving objects as well as advancing knowledge. One way to promote preservation and scholarship is to leave one's collection to a museum. The great size of Webb's holdings made that impossible, so in 1947 she created the Shelburne Museum, with a large and diverse collection that included weather-vanes, circus paraphernalia, trade signs, quilts, hunting decoys, and carriages, housed in a number of buildings and on the grounds. Many of these objects would certainly have been discarded had Webb not acquired them—this is preservation at its most basic. Because of her stature and diligence, and her ability to acquire so much, she bestowed legitimacy to a new area of collecting. As I have said, amassing groups of ephemeral items can yield important collections, offering endless possibilities for scholars of material culture. Museums and libraries—and other kinds of institutions with such holdings—prove to us that our cultural heritage is worth saving, regardless of how "high" or "low" the materials in the collection are.

There have probably been collectors for as long as there have been items to collect. The Renaissance inspired many members of ruling families, nobility, and the church to collect, and some of these collections were placed in the public domain. In Italy in particular, there are many municipal holdings that were started by prominent families centuries ago; some of their names are still familiar to us, such as Medici, Farnese, and Borghese. (In the United States, collecting is a more recent preoccupation, in part because most early European settlers to the United States had neither the time nor the means to be serious collectors.)

One type of collection, the *Wunderkammer*, also known as a Cabinet of Curiosities, was particularly popular in Europe in the sixteenth and seventeenth centuries (figure 5.2), though there are examples through the nineteenth century, including in the United States. They are well described by

Figure 5.2
Cabinet of Curiosity, Prague, 2016. Photo courtesy of Erica Ruscio, photographer

Florence Fearrington, a collector of books on the subject, who has curated several exhibitions about them.

A *Wunderkammer* is a room of wondrous things both natural and artificial (i.e., man-made), a chamber of objects noteworthy for their beauty, or their rarity; or their artistic or scholarly or monetary value. … they declined in the later eighteenth century, when a more systematic approach to the accumulation of natural and man-made objects developed. … *Wunderkammer* creators can be roughly divided into two main classes: (1) the nobility, and (2) physicians, apothecaries, and professional and amateur students of natural history.[11]

The collectors of these "rooms" and "cabinets" often categorized—and drew connections between and among—the items. They were seen as microcosms of the world. Some of these collections were intentionally gathered for research and study. In many cases, the collectors created catalogs, guides, and essays. (The Grolier exhibit that Fearrington curated included 137 such catalogs or guides.) *Wunderkammers* are precursors of museums, botanical gardens, and even circus freak shows.[12] Related to the Wunderkammer is the *Kunstkammer,* whose focus was on art rather than on natural history. As Fearrington's definition suggests, the original impetus of the Wunderkammer was to collect marvelous things gathered from nature. This attitude—wonder, marveling—made these collections of interest and were the reason for collecting in the first place. But human-made objects could elicit these same responses, and for most Wunderkammers, works of art found their way into these collections.

In 2014 the Museum of Fine Arts in Boston opened a new Kunstkammer Gallery, which includes precious objects, fine and decorative art, and exotic materials. The curators have filled the room with exquisite human-made objects and natural wonders, at the same time creating an intimate space that holds many treasures. iPads have been placed in the gallery to make it possible for visitors to see the fine details of the objects. Since such rooms were meant to show off the collecting prowess of the owner, the MFA has filled the room with fine objects including items that would have been popular during the height of the Kunstkammer craze, such as miniature paintings, furniture, clocks, automatons, and precious stones.[13]

* * *

Book collecting has a long and colorful history. Catalogs of private collections were often compiled by the collector, or by the collector's heirs or executors for estate sales. Tales of eccentric collectors abound, including accounts of voracious collectors whose desires outstrip their resources. Sir Thomas Phillipps (1792–1872) is most often cited for his excesses as a collector; he amassed a collection of some hundred thousand books and manuscripts and, despite his considerable wealth, he could not always pay for them (figure 5.3). His collection was sold in numerous lots from the late nineteenth- to the early twenty-first century. While he considered bequeathing his collection to the British government, he never got beyond preliminary discussions, in large part because of the unreasonable restrictions that he wished to place on access to his materials.

As much as Phillipps loved to amass materials, he believed that he had a preservation role to play by saving manuscripts from being broken apart— still a commonplace practice in his day (and not unknown today). One of his justifications for collecting so much was that he thought he could save that much more. As A. N. L. Munby and Nicolas Barker show, Phillipps's aim was "to have one copy of every book in the world."[14] Phillipps has been described as greedy, bigoted, and vain. Nonetheless, he amassed one of the greatest book collections in history.[15] As I have said, the collector's motives are not my focus in this chapter. But in Phillipps's case they are germane since he saw himself as a preservationist. It was anathema to him to see a butcher wrapping meat for his customers in leaves of vellum. For him—and ultimately for us—even scraps of manuscripts contained information. Our cultural heritage is that much richer through his preserving these historical items. Preservation takes many forms. In Phillipps's case, his collecting went far beyond the eccentric tendencies of some of the collectors depicted in Nicholas A. Basbanes's *A Gentle Madness: Bibliophiles, Biliomanes, and the Eternal Passion for Books.*

Sometimes a "gentle madness" crosses the line into another type of collector: the thief. Many items have been stolen from libraries and museums for the purpose of personal collecting. The most notorious book thief was Stephen C. Blumberg, who stole more than 23,000 books worth millions of dollars from libraries across the United States and Canada to enlarge his own specially created library in Ottumwa, Iowa. He was arrested in 1990. During his trial it came out that Blumberg sought to rescue and preserve the items that he stole. Basbanes included an essay on Blumberg in his book *Gentle Madness* because, although Blumberg was indeed a thief, his real aim was not financial gain but to have his own collection of remarkable, valuable books and manuscripts. In his case, stealing and building a great collection

Figure 5.3

"Portrait of Sir Thomas Phillipps," 1860. Permission of the Grolier Club, New York City. The collector poses with two items at the sale of the Rev. John Mitford's manuscripts on July 9, 1860: a tenth-century Horace (Phillipps MSS no. 15363), now in the Houghton Library, Harvard University; and a thirteenth-century copy of the gospels in Armenian [Phillipps MSS no. 15364], now in the Chester Beatty Collection, Dublin. The Grolier Club Library's copy of the photograph, which was taken shortly after the sale, is annotated by Sir Thomas with a description of the manuscripts.

were one and the same activity.[16] This is but one example of ethical lapses in "collecting."

With far fewer resources than Phillipps, Thomas Jefferson (1743–1826) similarly never seemed to have enough money for all the books he wished to acquire. Unlike Phillipps, who purchased books by the shelf and the shop, Jefferson was selective about his purchases. He sold his books—the largest personal collection in the United States—to the Library of Congress to replenish its collection of some three thousand volumes that were destroyed when the British burned the Capitol in 1814.[17] Jefferson sold his books for far less than he paid for them; he could not afford to give them away. True to his passion for collecting, after the sale he immediately began assembling a new library. In relation to preservation, he is best known for his observation, in a 1791 letter to Ebenezer Hazard, that America needed to have books in "a multiplication of copies, as shall place them beyond reach of accident."[18]

Jefferson's interest in book collecting never flagged. Toward the end of his life he was involved with the creation of the University of Virginia Library, selecting books and even designing the beautiful, stately rotunda that housed them (figure 5.4).[19] He was the rare collector who not only developed subject categories and a classification scheme for organizing his books, but also designed the spaces to house them. "I cannot live without books," Jefferson said. Sadly, his personal library did not outlive him for long: to pay off his considerable debts, his heirs sold his books at auction.

John Carter describes collectors in England who planned for the eventual donation of their collections to the British Museum, public libraries, or universities. This example of public spiritedness was also followed by less wealthy collectors such as David Garrick, the actor whose collection of plays went to the British Museum.[20] Carter rightly points out that some wealthy collectors are less disposed "to afford such public beneficence."[21] Their collections get passed down to heirs, or are purchased by dealers, or are auctioned off. This catch-and-release approach to collecting may be the preference of the collector, or the necessity of the heirs. But many institutions have been— and will continue to be—the beneficiaries of the generosity of collectors. Some of the greatest special collections in institutions were the result of the efforts of private collectors.

Great private collections have often been acquired by libraries directly (by gift or purchase from the collector) or indirectly through auctions or dealers. Library and museum curators have built superb collections item by

Figure 5.4
The Rotunda, University of Virginia (exterior and interior), 2016 and 2018. Photos
courtesy of Jane Penner, photographer

Figure 5.4 (continued)

item as well as by acquiring outside collections. Collection building is often a collaborative effort among institutions, private collectors, booksellers, auction houses, librarians, and others. Many museum and library directors have written about institutional collecting.[22]

With regard to the collectors considered in this chapter, Electra Havemeyer Webb started her own museum and Thomas Phillipps's books were sold at auction. One of Thomas Jefferson's libraries was sold to the Library of Congress while his final library was dispersed at auction. Jean Brown's collection of some six thousand items was sold to the J. Paul Getty Center in 1985. Like the other three collectors, she had a distinctive approach to her avocation.

Jean (Levy) Brown (1911–1994) was born in Brooklyn, New York, the daughter of a rare books dealer. She developed an early interest in art history, though she eventually became a librarian when she moved to Springfield, Massachusetts, in search of work. In 1936 she married Leonard Brown, an insurance agent from Springfield, and the couple made their home there.[23] (See figure 5.5.)[24] The Browns collected new art, purchasing Dadaist and

Figure 5.5
Jean Brown, n.d. Photo courtesy of Jonathan Brown

Surrealist art before they discovered the Fluxus movement, which included visual artists, poets, composers, and designers, and was an outgrowth of Dada and Surrealism. A noteworthy aspect of the Browns' collecting is that they acquired not just art objects, but also manifestos, newspapers, posters, and other ephemera that documented the movement (although there is a fine line as to what constitutes the art and what constitutes its documentation). Eventually Jean Brown also collected artists' books. She became friends with George Maciunas, the leader of Fluxus. The Getty finding aid for the collection describes Jean Brown's approach:

Brown's primary goal was to assemble a study collection. She acquired comprehensively on the topics mentioned above. This included standing orders with some small presses to acquire all of their output. Her early appreciation of books lead [sic] naturally to an interest in artists' books. If an artist's work interested her she asked the artist to create a book for her archive. In the early 1970s, her son Jon sent notices about the archive to every art history graduate program. Scholars and graduate students with valid research interests were invited to use the collection.[25]

Leonard Brown died in 1971 and Jean moved into a Shaker seed house in Tyringham, Massachusetts. The house, itself a work of art, became a gathering place for Fluxus artists. She developed close relationships with artists whose work she collected, including George Maciunas, Dick Higgins, Ken Friedman, and Peter Frank. Some of them, in Brown's upstairs workroom, created works for the archive. It is tempting to think of this Shaker dwelling as seeding Fluxus creativity.

Jean and Leonard's son, Jonathan Brown, gives a vivid description of his parents' collecting. They were early collectors, by Springfield standards, of contemporary art. Their passion for it led them to the art galleries of New York City where they met and befriended many contemporary artists. After acquiring a copy of Robert Motherwell's *The Dada Painters and Poets*, they decided to collect the documents and publications of Dada and Surrealism.[26] Brown describes his parents as engaged and engaging collectors. One also gains that impression from studying their archive at the Getty.

There have been a number of challenges associated with preserving Fluxus materials: the works of art might be on poor or fragile materials; mail art (creating small scale art works and sending them through mail) may have been damaged en route to its destination; items of "new" media become obsolete. A more subtle problem is the complexity of preserving performative art. If an event was not filmed or taped, then we are reliant on posters, programs, or flyers that advertised those performances. Yet these may not accurately capture what actually happened at the event. For example, artists who never performed might be listed on a program, or those not listed might have performed. While this is true for any live event, it is particularly true given the improvisational nature of Fluxus art. The role of people who participated in Fluxus is as important as is hearing from those who actually attended the events.[27] While those who participated in or observed Fluxus happenings can still be interviewed, the media used to record the interviews will also need to be preserved.

There are also preservation and conservation challenges with the Fluxus art works. A partial listing of the items at the Getty reveals why: there are event scores in mixed media, and idea boxes, games, sound recordings (including cassette tapes), moving images, and video recordings. Each of these media has its own requirements for proper storage and handling. (Some approaches to the challenges of new media are considered in chapter 7, "What Are We Really Trying to Preserve?")

Jean Brown played a performative role in the Fluxus movement as mentor, hostess, and collector. The Getty collection includes clips of Brown in her home with Fluxus objects. Her collection is particularly rich for students of Fluxus, because it invites us to be part of the movement as well. I dwell on Brown's collection here because it reveals some of the challenges we as preservers of cultural heritage face, and issues touched on throughout this book: the nature of collections and their collectors; the vicissitudes of the world; the tremendously varying kinds of things collected; institutional challenges; and much more. In its own way, collecting is a microcosm of the larger world of preservation.

*   *   *

In Tanselle's definition of collecting, there is one gap that limits a full consideration of preservation and collecting—the exclusion of "intangible things." He explains that people may collect memories for which there can be no physical possession, and that there is a difference between "an internal repertoire of ideas and an external grouping of tangible materials."[28] However, memories become stories, rituals, and traditions, and there are aspects of the intangible that *can* be collected since these can be recorded in analog media.

For example, a storyteller is a collector of stories. In oral cultures, stories (or histories) may be passed from generation to generation and preserved only in their transmission. In fact, family stories are passed down orally between generations even in cultures that record their histories in a tangible form. Preservation ends when the last person in a family or tribe dies—unless someone is around to record that history. In such cases, the heritage becomes tangible—if it is preserved.

There are inherent differences between oral and written cultures. The oral tradition is performative, so each rendering of a story is different from the previous version. Once a story is recorded or written down, it becomes fixed; what is preserved represents only one version, or performance, of the story. Thus it is not possible to determine what the "original" version is, if the concept of "original" even makes sense. What may be lost on the listener (or reader of the transcript) is the knowledge that there are inherent variations in oral transmission.[29]

Recordings of oral traditions are collected by institutions of many kinds. In some circumstances, restrictions may need to be placed on use of and access to the recordings, out of respect for the customs of a particular tribe, for example, or because the recordings were made illegally, and therefore some people think they should be destroyed. There are ethical issues that must be addressed when recordings are made without (or even with) the knowledge of the person or people being recorded. The point is that in some cases, people may wish to have their histories or customs preserved for posterity. The preservation of a recording makes it possible to preserve intangible culture in a tangible medium—though in most cases only one version is being preserved, and the medium is not a 100 percent faithful record of the intangible culture.

UNESCO adopted the Convention for the Safeguarding of the Intangible Cultural Heritage in 2003. It states: "Cultural heritage does not end at monuments and collections of objects. It also includes traditions or living expressions inherited from our ancestors and passed on to our descendants, such as oral traditions, performing arts, social practices, rituals, festive events, knowledge and practices concerning nature and the universe or the knowledge and skills to produce traditional crafts."[30]

Furthermore, the convention states that intangible cultural heritage becomes heritage only when the communities, groups, or individuals who "create, maintain and transmit it"[31] recognize it as such. This community-based perspective means that there are many people who have a right to weigh in on their cultural heritage. A diversity of voices leads to many points of view. These perspectives relate directly to collecting and preservation.

Who gets to decide what is collected? And who determines what is preserved? These key questions are part of a much larger picture. There are also ancillary questions: who will do the actual collecting? Who will store (i.e., preserve) it all? At whose expense? For how long? To what end? Who will [be permitted to] use a collection? How will it be used? And so on.

Intangible culture becomes tangible if it is recorded in some way. As mentioned earlier, the intangible can become tangible. But the tangible can also become intangible. For example, anything created can be destroyed; and something that is created or presented online can disappear. We usually refer to analog or digital objects that may disappear as ephemeral, but items intended to be permanent can also disappear. Collecting may or may not be a hedge against disappearance.

Collecting is the accumulation of tangible *and intangible* things. It may not ensure preservation, but it is the necessary first step. We are what we collect—and preserve.

This chapter has looked at collecting, with a focus on essentially small things: books and manuscripts. Though a single book may be seen as a monument (the Book of Kells, Gutenberg's forty-two-line bible, the First Folio of Shakespeare's works), and although even a single document may also be called a monument (the Magna Carta, the Declaration of Independence), we do not generally think of books and manuscripts as monumental. But collections certainly are. One or even ten items could have great research value. But many libraries have amassed large collections in particular subjects, be it first editions, journals, children's books, publications on arctic exploration, or any other area—clearly these are monumental resources. Hence, collections can have world-class possibilities for scholars, and their preservation is key to our accessing cultural heritage on a monumental scale.

6 Worth Dying For? Richard Nickel and Historic Preservation in Chicago

The photography and salvaging of ornament is a final preservation measure for the benefit of historians and is in no way a substitute for the whole building. Architecture is evaluated by form, proportion, placement of parts and special qualities. The aesthetic experiencing of a whole building is not possible with photographs or pieces. They are symbols.

—Richard Cahan, *They All Fall Down* (1994)[1]

Historical preservation is—among other things—the effort to save buildings, groups of buildings, districts, neighborhoods, or even entire towns. Efforts may focus on the built environment alone, on the natural environment, or both. Preserving buildings dates back to the earliest human settlements in China, Pakistan, and Mesopotamia where the structures were built from such materials as mud brick that required constant repair and rebuilding. In that sense, according to John H. Stubbs,

preservation-minded actions across the millennia can be considered the genesis of today's architectural conservation ethos for two reasons: 1. The continuous use of old buildings forced the concept of extended use, adaptive use, and maintenance, thus making preservation for many structures inevitable. 2. Such maintenance and preservation supported the survival of an increasing number of historic buildings and sites.[2]

An early example of restoration can be seen in the back of the seated figure of Ramses II at the entrance of the Luxor Temple in Egypt.[3] In Rome toward the end of the Western Roman Empire, Emperor Majorian issued an edict, *Novella Maioriani 4, De aedificiis pubblicis*, in C.E. 459 or 460 to end the destruction of the empire's monuments. The edict was aimed at builders who plundered monuments rather than travel to distant quarries for their materials. Builders were not the only plunderers, however: by that time Rome was under constant attack and its architectural treasures were being repeatedly looted or destroyed by invaders.[4]

This chapter explores the efforts of Richard Nickel (1928–1972) to preserve buildings that were designed by the architectural firm of Adler & Sullivan (its principals Dankmar Adler and Louis Sullivan) and other leading
architects of the so-called Chicago School, such as Henry Hobson (H. H.)
Richardson, William Le Baron Jenney, Daniel Burnham, John Wellborn
Root, John Holabird, and Martin Roche.[5] While Nickel was not able to
prevent the tearing down of some of Sullivan's buildings in Chicago, he
brought publicity to their plight and documented them before and during
their demolition. Sullivan refined the use of steel girders in the skyscrapers
that he designed. He advocated for a form-follows-function approach to
designing buildings and the design elements of his buildings were beautiful. (And, as we shall see, Nickel salvaged many of his cornices and other
decorative elements.) For these reasons he is now considered, along with
H. H. Richardson and Frank Lloyd Wright (whom he mentored), to be one
of the fathers of American architecture. Yet when Nickel was a student, Sullivan's work was largely ignored. The documentation of Sullivan's buildings
in particular became Nickel's life's work.

Nickel made three significant contributions to historic preservation: he
systematically photo-documented buildings; he salvaged ornamentation
and other features of a building whenever it was possible; and he gathered
and preserved as many records as he could that related to an endangered
building. These aspects of his work will be considered in this chapter.

Today, Nickel's talent as a photographer, and the fruits of his meticulous
attention to archiving his work, can be seen in the Richard Nickel Archive
at the Ryerson and Burnham Libraries of the Art Institute of Chicago—the
central focus of which is the architecture of Dankmar Adler and Louis Sullivan.[6] This archive can be considered Nickel's fourth contribution to historic
preservation: it contains more than fifteen thousand photographs, contact
sheets, architectural drawings and reproductions, pieces of correspondence,
and other items relevant to architectural history.[7] The archive documents a
key chapter in Chicago architecture and is a rich resource for scholars. Nickel's efforts to salvage portions of the Chicago Stock Exchange—built in 1893–
1894 and demolished in 1972—resulted in a partial reconstruction of the
Trading Room, also at the Art Institute.[8] Nickel gave his life to that project:
toward the end of the demolition of the Stock Exchange building he fell to
his death when the partially demolished structure beneath him gave way. His
corpse was not found until nearly a month later, "two floors directly beneath

the Trading Room in the building's sub-basement."[9] What was to have been Nickel's last salvage effort for one building instead tragically became the final salvage effort of a brilliant career (see figures 6.1 and 6.3).

Historic preservation in Chicago is another theme of this chapter. The Chicago Architecture Foundation describes the city this way: "Chicago has long been a laboratory for architectural innovation and experimentation."[10] But historic preservation in the city has been more complex than that claim implies. Many of its landmark buildings—by revered architects like Sullivan and H. H. Richardson—were torn down, including, ironically, the Home Insurance Building (1884–1885), the nation's first steel-frame skyscraper, designed by William LeBaron Jenney. It is the type of structure for which the city is best known. Thus Chicago's architectural legacy contains countless episodes of demolition as well as preservation. Just as a collection development plan in a library must include plans for weeding and deaccession, so must a survey of the preservation of a city's architecture explain the

Figure 6.1
"Chicago Stock Exchange Building, Chicago, IL. Exterior Detail." Richard Nickel, photographer. File # 201006_110816–028. Richard Nickel Archive, Ryerson and Burnham Archives. Permission of the Art Institute of Chicago

Figure 6.2
"Chicago Stock Exchange Building, Chicago, IL. Removal of Terra-Cotta Cornice
Ornament." File # 201006_110816–009. Richard Nickel Archive, Ryerson and Burnham
Archives. Permission of the Art Institute of Chicago

loss of any of its architectural sites. Indeed, the whole notion of *preservation*
must be considered in the context of that loss. We would not need a notion
of preservation if there were no forces in the world that make preservation
desirable—forces of decay and destruction and loss, whatever the sources of
these forces are—whether natural or human-made.

There are as many rationalizations for the destruction of buildings as there
are for preservation: buildings may no longer serve their purpose; they are in
disrepair and would cost too much money to repair and restore; they are taking up valuable land that could serve better ends (read "be more profitable");

Figure 6.3
"Detail, the Trading Room as Reconstructed in the Art Institute of Chicago, 1976–77, Chicago, IL." File # 201006_120808–017. Richard Nickel Archive, Ryerson and Burnham Archives. Permission of Bob Thaw, photographer

or they are eyesores (usually a justification based on "economic aesthetics" in which the buildings are not so much ugly as they are impediments to a "better" use of the land). Often the decision to tear down a building is made without public input. Thus, over the past century means have been established for preserving buildings for the public and in national or local "trust." But also, other means have been used to get rid of what we in hindsight see as important buildings.

The complexities and contradictions of historic preservation in Chicago make it a rich city to study. Chicago prides itself on its architecture and yet, as is true in many other cities, significant buildings and neighborhoods continue to be torn down. Indeed, Chicago exemplifies the many issues of historic preservation that have played out everywhere.

I have selected Nickel as a lens through which to consider historic preservation and its role in "monumental preservation." According to Max Page and Randall Mason, the history of American historic preservation has been dominated by examples of influential people who saved buildings (e.g., Ann Pamela Cunningham and Mount Vernon, Henry Ford and Greenfield Village, William Sumner Appleton and the Society for the Preservation of New England Antiquities), or the preservation of "important" buildings such as Grand Central Station in New York City. Page and Mason have characterized these efforts as "the great 'saves.'"[11] Their 2004 book *Giving Preservation a History* has broadened that history by including scholarship about the social history of preservation. Their approach has placed historic preservation into a wider social and cultural context by exploring the motivations for preservation, the marketing of it, the role of the public, economic development, class issues, grassroots approaches (e.g., who gets displaced as a result of the demolition of buildings), and so on. Page and Mason have also considered the often limited vision of historic preservation that until recently did little to document African-American, Native American, and immigrant experiences. Put another way, Robert E. Stipe has described the movement in historic preservation since around 1980 away from being focused on saving "stuff" to considering the impact of preservation on people.[12]

The aim of this chapter is not to build on their important work. Rather, it is to put historic preservation into the context of preservation more generally. Page and Mason see historic preservation as associated "with other history industries."[13] But historic preservation is also part of a larger undertaking to preserve all heritage: documents, art, the built and natural environments, customs, languages, and nature. Preservationists in all heritage fields can learn from historic preservation with its dependence on diverse relationships: the tensions between individual property rights, developers, and the public good; between business and the national trust; and the often divergent perspectives about the past and interpretations of it. Historic preservation advocates must engage with builders and real estate developers, financial investors, lobbyists, politicians and community leaders, and often the general public. Economic development, tourism, tax credits, building codes, and laws are all components of historic preservation. While preservation in general involves a variety of constituencies, no other preservation specialization is so dependent on complex relationships. Thus the preservation of documents in an archive, books in a library, or objects in a museum is far less dependent on external relationships. In short, historic preservation is at the heart of community building and development.

A study of Richard Nickel fits into new scholarship on historic preservation. That is, by studying Nickel we can learn about the process of historic preservation. Unlike the mostly wealthy or well-connected preservationists who came before him, Nickel was from a working-class, not an elite, background. He was largely unsuccessful at saving buildings yet he greatly advanced our knowledge about Louis Sullivan through his tireless, even obsessive devotion to documenting that architect's work. With this kind of documentation, Nickel laid the groundwork for others. If, as Page and Mason suggest, historic "preservation [is] a *process* rather than ... a set of *results*,"[14] Nickel's work can perhaps be seen as a prelude to this approach. While Nickel focused on results—saving the buildings of Adler and Sullivan and others—his legacy is also tied to his processes. (Among his many jobs, he photographed buildings for the Historic American Buildings Survey, a federal program started in 1933.)[15]

Historic preservation is also about the long—or short—reach of history and memory. Richard Nickel sought to preserve particular Chicago buildings, and his story is part of a larger narrative of the often-conflicting approaches to historic preservation in Chicago. As Daniel Bluestone has described it in a piece aptly titled "Preservation and Destruction in Chicago," "The idea that some history was no longer historic or worthy of chronicling or saving was common in Chicago."[16] A wonderful example of Bluestone's observation can be seen on a plaque on the outside of the main building of the Newberry Library, which was designed by Henry Ives Cobb in 1893 (see figure 6.4).

Today, the house would probably be preserved. But a hundred years ago the text on a plaque documenting that demolition could be written without irony. (In fact, the act of memorializing the house with a plaque might have been seen as a preservation act.) Yet even before the "Great Fire," Chicagoans had established the Chicago Historical Society in 1856, when the city was only twenty-five years old. A century later, the Chicago City Council established the Commission on Chicago Landmarks during the peak of an urban renewal movement that resulted in the leveling of thousands of buildings—a number of which were historically significant. Historic preservation in Chicago reflects the city's identity politics, for better or for worse. Sadly, residents of poor neighborhoods did not have the political clout to preserve their buildings. But even buildings in higher-end neighborhoods were sometimes razed in the name of progress (read "fiscal gain").

Figure 6.4
Plaque at the Newberry Library. Permission of Linda M. Chan, photographer

Daniel Bluestone has written about the destruction of the "Mecca," which was a hotel during the Columbian Exposition of 1893 (see figure 6.5). Designed in 1891 by Willoughby J. Edbrooke and Franklin Pierce Burnham, the building was known for its use of natural light and interior atrium design. And while the elegant physical properties of the building have been noted, just as significant—but often overlooked—is that the building stood in an African-American neighborhood. As I have just noted in my previous paragraph, those living in the neighborhood did not have the power to save their building the way that wealthy people might have. Bluestone says,

Over time, race intersected with urban space to alter the history and fragment public perceptions of the Mecca. These changing perceptions stood at the center of a decade-long preservation struggle. Although the early preservation movement generally adopted prevailing notions of Chicago School aesthetics as its point of departure, the Mecca campaign emphasized housing and neighborhood. In place of an aesthetic model for preservation efforts, the Mecca's story recovers a series of alternative priorities.[17]

The Mecca was torn down in 1952 as part of the expansion of the Illinois Institute of Technology (now also known as Illinois Tech and IIT). The

Figure 6.5
The Mecca Hotel. Photo courtesy of the City of Chicago Department of Cultural Affairs and Special Events, from the *Mecca Flat Blues* exhibition at the Chicago Cultural Center, February 15–May 25, 2014.

building did not fit into the new aesthetic of IIT, so campus planning did not include a proposal for restoring the Mecca. In its place, Ludwig Mies van der Rohe's S. R. Crown Hall was built. The building is considered one of van der Rohe's masterpieces, and today it is on the U.S. National Register of Historic Places and is a listed U.S. National Historic Landmark and Chicago Landmark. But of course, its future as a landmark could not have been anticipated then, nor does a strictly aesthetic appreciation of Crown Hall in itself justify the tearing down of the Mecca. Had the Mecca been preserved and restored, it would most likely have been used as a campus building. The bottom line, however, is that an African-American community was displaced to make way for an expanding university.

The demolition of deteriorating buildings that housed the poor would take place on a large scale under Mayor Richard J. Daley (mayor from 1955 to 1976) in his quest for *urban renewal*, a term that refers to the combination of federal, state, local, and private initiatives that were put in place to revitalize cities.[18] In Chicago this led to slum clearance, made way for urban "revitalization," and stemmed the tide of white flight to the suburbs—which would have diminished the city's tax base. Land taken by the city through eminent domain was turned over to private developers who renewed Chicago's downtown area, known as the Loop, as well as many other parts of the city.

While the impetus of such "renewal" is almost always economic, the negative impact this has on poor people's lives is monumental. And such impact has no negative effect on those who stand to profit most from the demolition. Two issues dominate. First, in the name of improvement, the city must begin with destruction, regardless of the aesthetics or historicity of that which is being destroyed—and, again, regardless of the human impact. Second, to justify such cataclysmic actions (tearing down monumental buildings is cataclysmic on a physical level, and it is cataclysmic to those who are being displaced), city developers use the euphemism *renewal* with the implication that all will benefit from the demolition. Perhaps in the long run the city does benefit, but at what cost? Certainly, there is a cost to being displaced. But in the field of historic preservation, there are other costs to be evaluated, such as the loss of historic buildings, or the sociology of the neighborhoods. Change can often be justified with euphemisms and rhetoric, but is change always for the better? Is it always justified?

Also, city planners, completely disregarding the sociology of destroying old (and possibly historic) buildings, and thinking only of the fiscal implications for the city (and possibly also for themselves), can mask their depredations by using the euphemism *revitalization*. Where is the revitalization? At whose expense does it occur? This latter question, of course, does not figure into the discourse of those who stand to gain from the destruction of our cultural heritage. And the people who are the "victims" of revitalization (usually the displaced poor) have little or no say in the matter at all.

While social justice was not the focus of Richard Nickel's work, urban renewal impacted it. Buildings by H. H. Richardson, Louis Sullivan, and others that were in "blighted" neighborhoods were torn down. Willoughby J. Edbrooke and Franklin Pierce Burnham's Mecca was not the only building leveled by IIT as the campus expanded from 7 to 110 acres.[19] Hirsch describes the expansion of IIT, and later other schools, as part of "a twenty-year burst of activity [that] nearly doubled downtown office space; the federal government, Cook County, and the city of Chicago [which] each added massive administrative centers."[20]

While efforts were being made to save the Mecca, Richard Nickel was serving in the Korean War. Nickel had enrolled at IIT in 1948 and resumed his studies there after his return from Korea. His master's thesis, "Louis Sullivan's Architectural Ornamentation," was written in 1954, not long after the demolition of the Mecca. However, Nickel's focus was, with few exceptions, on Sullivan, whose work he systematically documented for the rest of his short life.

Nickel's Contributions

Richard Stanley Nickel was born in Chicago, the son of first-generation Americans.[21] His father, John, was a driver for the *Polish Daily News*; he Americanized his surname Nikiel to Nickel. Richard Nickel served in the army and then enrolled at the Institute of Design, now IIT. He married in 1950, served in the Korean War, divorced, and moved back in with his parents—where he remained for the rest of his life. He purchased a house in Logan Square but never lived there; however it was a base for his salvage work.

The turning point in his life came when he took a class in the 1950s with Aaron Siskind (1903–1991), who introduced him to architectural photography and the buildings of Dankmar Adler and Louis Sullivan. Nickel's master's thesis project was to photo-document Sullivan's buildings. In his research, Nickel discovered many Sullivan buildings that were not known to be Sullivan's work. He continued this work for the rest of his life, tracking down Sullivan's many previously unascribed buildings. By the time he died, he had identified and recorded dozens of Sullivan edifices—many just ahead of the wrecking ball. He is buried in Chicago's Graceland cemetery, not far from Louis Sullivan (figure 6.6).

Historic preservation gained momentum in Chicago in the 1960s—the peak period of Nickel's preservation activities. The interest was in response to the massive demolition of nearly six thousand buildings from 1957 to 1960 alone,[22] construction of expressways that destroyed middle-class and lower middle-class neighborhoods, and the demand for increased office space in Chicago's Loop.[23] Nickel's efforts to keep up with the demolition of Sullivan buildings were hampered by the pace of the city's development. In 1966, the Chicago Architecture Foundation was established as part of an effort to purchase and preserve Glessner House, the masterpiece residence designed by H. H. Richardson, and today one of Chicago's treasured historic buildings. A short time later, the Commission on Chicago Landmarks (CCL) was founded. The commission "researches the background of properties or districts proposed for landmark status and recommends approval to the Chicago City Council, which then votes on granting such status."[24]

Another significant development of the period was the passage by Congress of the National Historic Preservation Act (NHPA) of 1966 that created

Figure 6.6
"Nickel with Large Format Camera, Chicago, IL." File # 201006_161213–001. Richard Nickel Archive, Ryerson and Burnham Archives. Permission of the Art Institute of Chicago

a national system built on federal, state, and local government partnerships. Urban renewal activities throughout the United States provided impetus for the NHPA. It expanded the reach of earlier programs such as the 1935 Historic Sites Act and the National Trust for Historic Preservation (1949). These earlier acts focused on nationally significant places and properties. The 1966 act was an attempt to preserve places of significance to local communities. New financial incentives could be put in place to assist preservation initiatives.

Figure 6.7
"Adler & Sullivan, Building for Richard Knisely, 1883." Richard Nickel, photographer. File # 201006_110815–072. Ryerson and Burnham Archives. Permission of the Art Institute of Chicago

Yet even with the existence of preservation organizations and laws, the fate of individual buildings is also dependent on economic considerations. This is where Richard Nickel often met insurmountable challenges. Louis Sullivan designed many commercial buildings in Chicago. The owners of the land could tear them down and build larger and more modern buildings to replace them. Not only is historic preservation expensive, but restored buildings might result in less space than a new building could offer. Property owners wanted to maximize their profits, and there were plenty to be made in downtown Chicago real estate in the 1960s. In other cases, Sullivan buildings in poor neighborhoods had been neglected for so long that it was no longer practicable to save them (figure 6.7).

* * *

While preservation activities have occurred for a long time, the move toward a coordinated approach to what we now call "historic preservation" (also referred to as *architectural conservation*) has been gradual, with some significant developments along the way, particularly in the nineteenth and twentieth centuries, such as the protection of parks and other open spaces, the city beautification movement, and so on, which are discussed in chapter 9, "Sustainable Preservation."

In 1931, the Athens Charter created an international network for preservation, including the creation of the International Council of Museums (ICOM), and the International Centre for the Study of the Preservation and Restoration of Cultural Property. Thirty-three years later the International Charter for the Conservation and Restoration of Monuments and Sites of 1964, usually referred to as the Venice Charter, furthered the aims of the Athens Charter. Its aims are defined in the first three articles:

Article 1. The concept of a historic monument embraces not only the single architectural work but also the urban or rural setting in which is found the evidence of a particular civilization, a significant development or a historic event. This applies not only to great works of art but also to more modest works of the past which have acquired cultural significance with the passing of time.

Article 2. The conservation and restoration of monuments must have recourse to all the sciences and techniques which can contribute to the study and safeguarding of the architectural heritage.

Article 3. The intention in conserving and restoring monuments is to safeguard them no less as works of art than as historical evidence.[25]

Historic preservation encompasses particular buildings, the settings in which buildings are created, and, though not explicitly stated in the Venice Charter, sites that are connected to social histories. The charter does not specifically address either the concept of site that applies to historic landscapes and gardens, or the social and financial issues that inhere in preservation. And while the Venice Charter gives an international framework to the protection of buildings, historic preservation is also necessarily tied to local governmental and private interests. Also, historic preservation takes into account the human element: who built the structures? Who owned them? Who is impacted by their destruction? Thus it is almost always a challenge to create a sustained, coordinated approach to historic preservation.

*   *   *

Richard Nickel recognized that attempts to preserve Sullivan's commercial gems would mostly be futile. Rather than give up the fight, he took every possible preservation action: appealing to the buildings' owners or to Chicago's

Mayor Daley, attending public planning meetings, and organizing protests (for example, to preserve the Garrick Theater). When those strategies failed, he photo-documented the buildings thoroughly from the inside out. His detailed photographs captured much of the character of the buildings. He also salvaged as many of the decorative elements as he could. He sold such items to Southern Illinois University, Edwardsville, and the Art Institute of Chicago, as well as some other institutions.

Nickel's other preservation strategy was to write a comprehensive work on Adler and Sullivan. He labored for years to write a book, which was completed only in 2010, by his colleagues and friends and the Richard Nickel Committee nearly forty years after his death.[26] Here is an example of process over product. Richard Cahan amply documents Nickel's struggles with the writing of his book (notably his perfectionism and insecurity) as well as the fact that in the late 1950s and throughout the 1960s, Adler and Sullivan buildings were constantly being torn down. Nickel could barely keep up with the demands of thoroughly documenting so many buildings—he took some fifteen thousand photographs[27]—and also salvaging pieces of them. Add his constant struggle to support himself, and the product of his labors, his book, nearly lost out to the process of documentation.

Nickel's photographs are formalistic, sometimes elegiac, and beautiful. There is also an implied sadness: these majestic buildings cannot defend themselves against people. (In fact, an oft-repeated quote by Nickel is "Great architecture has only two natural enemies: water and stupid men."[28]) Yet one can also glean something more from the photographs (see figure 6.7). This image of the Knisely Building is shot straight on, with a long depth of field, and it derives its power in part because of the inclusion of the wrecked cars in the foreground. The building, soon to be demolished, seems already to have "fallen victim" to the company it now keeps. Yet the integrity of the building is not compromised by its next-door neighbor, the wrecking yard. There is more than a little irony in the photo: in the late 1940s and 1950s, many slums in Chicago were bulldozed. Mayor Daley and others made it their goal to eliminate them, almost as if they were a contagious disease. Daley said, "We must have slum clearance. While we are clearing the slums, we must prevent the blight from spreading into the other neighborhoods."[29] With or without slums, the wrecked cars depict their own blight. Yet it is likely that this wrecking yard was around well after the Knisely Building was demolished. But the very presence of the wrecking yard in the photo shows the precariousness of the monumental building. Wrecked cars are not gathered in lots in up-scale neighborhoods.

At the beginning of this chapter I identified Nickel's three contributions to historic preservation: saving as many records as possible relating to an endangered building (e.g., blueprints, architectural renderings, historical photos, newspaper and magazine articles); photo-documenting the building before and during its demolition; and systematically salvaging ornamentation and other features of the building. Nickel's painstaking work meant that buildings that might have been demolished almost without notice could be documented in minute detail. Nickel's methods included meticulous photo-documentation, the rescuing of floor plans, interviews with tenants of the buildings, and salvaging—in the days before it was popular or lucrative. Nickel viewed himself as a documentarian. His ability to convince Southern Illinois University, Edwardsville, and other institutions and organizations that they should purchase remains and ornaments before it was common for museums to do so showed much foresight. (Until 1961, Nickel had stored salvaged pieces at Chicago's Navy Pier because no single institution wanted everything that he had.[30])

Nickel's almost singular focus on Louis Sullivan resulted in exhaustive preservation efforts of the once-neglected architect, and his bearing witness to destruction impacted historic preservation—but ultimately killed him. As I noted earlier, he died while he was salvaging pieces in the partially demolished Stock Exchange building. The structure gave way and he plunged to its basement. Ironically, however, Nickel's occupation might have killed him eventually: the autopsy revealed that although he had been crushed to death, he already had pulmonary emphysema and chronic bronchitis. Nickel was a nonsmoker, but his condition was probably caused by the dust and other materials that he inhaled during the many years he spent documenting buildings in the process of demolition.[31]

In "Richard Nickel's Photography: Preserving Ornament in Architecture," Sarah Rogers Morris holds that Nickel's efforts to document the decorative schemas and salvage ornamental fragments of Louis Sullivan's buildings provide a continuous link to modern design. Without Nickel's efforts, Sullivan's work might have been lost to a newer generation of architects who largely disregarded ornament. (In fact one of those architects, Mies van der Rohe, taught at IIT, Nickel's alma mater.) Hugh Morrison[32] and Nickel himself

Figure 6.8
"H. H. Richardson's Glessner House, 1887." Pastel by Jane Steele, 1983. Photo: Paul Gulla; image courtesy of Suzy Steele Born

believed that Sullivan's work anticipated modernist designers. After all, Adler and Sullivan pioneered skyscrapers and skeleton construction. When Nickel was trying to save Sullivan's Garrick Theater, architect Le Corbusier wrote to Mayor Daley urging him to preserve the building, further emphasizing Sullivan's links to modernism.[33] Thus, Morris notes, Nickel "restored Sullivan's conception of a fluid, continuous program of architecture and ornament."[34]

Politics and money have often made it difficult—or even impossible—to save important buildings. In Chicago alone, without the valiant efforts of Nickel and others, Frank Lloyd Wright's Robie House, H. H. Richardson's John J. Glessner House (his last surviving building in Chicago), and perhaps all of Louis Sullivan's buildings might have been torn down (figure 6.8).

A more recent Chicago tear-down shows that even in a city that today prides itself on its architecture, the push-pull of private and governmental interests may lead to the destruction of architectural gems. Bertrand Goldberg's Prentice Women's Hospital (1975) at Northwestern University was torn down in 2013–2014 to make room for a new building. Goldberg

(1913–1997), a Chicago architect, is best known for his corncob-shaped Marina Towers (see figures 6.9 and 6.10). He studied with Mies Van der Rohe, and his buildings span six decades. Several preservation organizations tried to save the hospital building: Save Prentice Coalition, DOCOMOMO, the National Trust for Historic Preservation, and Preservation Chicago. However, the groups ultimately lost their battle with Northwestern University. Northwestern could justify tearing down Prentice because a new research facility would be "aimed at attacking heart disease, cancer, and children's diseases," which, apparently, it could not accomplish by building on its nearby empty lots. Northwestern demonstrated that rhetoric and an aggressive PR campaign can be important in making the case to destroy a building. Once again language was a tool for justifying the loss of historic monuments. The emotional tug of saving our children and attacking cancer and heart disease won out. In response to Northwestern's demolition of Prentice Women's Hospital, *ArchDaily* published this commentary:

Following the extensive preservation battle over Bertrand Goldberg's iconic Prentice Women's Hospital, the Chicago landmark was demolished a few months ago to pave the way for Perkins+Will's new Biomedical Research Building for the Feinberg School of Medicine. The four year preservation struggle was marked by repeated appeals to the Commission on Chicago Landmarks and Mayor Rahm Emanuel with attempts to place the building on historic registers, proposals to adapt it for modern use, and design competitions to gain public opinion on the future of the building. Ultimately, the outpouring of global support by architects and preservationists to save Prentice fell short of the political agenda of progress, prioritizing future development over preserving the city's past.

In the wake of the loss of this icon, the National Trust for Historic Preservation has released a time-lapse video documenting the demolition process of Prentice from start to finish. This incredible footage memorializes the one-of-a-kind building so although the new Biomedical Research Building will soon take its place, a piece of its predecessor will always be remembered.[35]

The video mentioned in the preceding commentary is a twenty-first-century echo of Nickel's still images. But unlike Nickel's work, this video does not stand on its own as an artistic achievement; it merely records the demolition.

It is worth repeating one key point: the architectural (and cultural heritage) importance of the Prentice Women's Hospital is beyond question. But developers played on the emotions of city officials (who allowed the destruction) and the public (who witnessed it) by conjuring up pictures of sick children and people with cancer and heart disease. As I have noted

Figure 6.9
"Marina City, Chicago, IL." File # 200203.081229–310. Bertrand Goldberg Archive,
Ryerson and Burnham Archives. Permission of the Art Institute of Chicago

Figure 6.10
"Northwestern Memorial Hospital, Exterior View of Prentice Women's Hospital and
Maternity Center, Chicago, IL." File # 200203.081229–421. Ryerson and Burnham
Archives. Permission of the Art Institute of Chicago

elsewhere in this book, preservation has an emotional component that, if properly manipulated, can have disastrous (or excellent) effects. The strategic use of language can affect how our cultural monuments live or die.

* * *

Historic buildings are not the only structures that continue to be torn down. In 2000, Chicago's Maxwell Street Market was pulled down (to make way for upscale condos and restaurants) and moved to Desplaines Street. It has been described as the "birthplace of the urban electric blues, a place where Muddy Waters, Robert Nighthawk, Little Walter, Hound Dog Taylor, and dozens of other musicians found ready audiences and turned an African-American musical form into a foundation of modern popular music."[36] The Market, dating back to the 1880s, served generations of Chicagoans. Can its gritty and colorful characteristics be preserved?

Chicago architectural photographer and blogger Lee Bey is the spiritual descendent of Richard Nickel. Lee Bey Associates is a research-based consulting practice that works with real estate developers, community agencies, and architecture and urban planning firms in governmental affairs and civic engagement. He produces and hosts the *Architecture 360* podcast, which focuses on architecture, preservation, and urban planning in Chicago. In keeping with the changing focus of historic preservation described by Page, Mason, and Stipe, quoted earlier, Bey focuses not just on individual buildings but also on urban planning, education, affordable housing, and economic development in Chicago's Loop.

Lee Bay Associates represents not a new effort at preservation, but the ongoing process of preservationists to save historic properties. Although architectural gems like the Prentice Women's Hospital continue to be torn down, there has been an ongoing evolution in historic preservation. Rather than mostly focusing on individual buildings, preservationists today focus on communities as well. There is a greater public awareness of historic preservation than there was when Richard Nickel was working, and alliances now exist among many stakeholders. It will never be possible to save everything worthy of preservation; big business interests will continue to prevail using the tactics described here. Monumental, indeed, will be continued efforts at preservation.

Coda

Studs Terkel (1912–2008), Chicago's plucky chronicler, exercised some poetic license when he described Nickel this way:

Anyway, when Richard Nickel heard they were tearing down the Stock Exchange [b]uilding, one of the jewels in Sullivan's crown…he went down there to save whatever he could. A piece of this, a piece of that. He was a blessed scavenger.

I guess he was down there one day while the wreckers were still at work. He couldn't wait. He was afraid they'd cart the stuff off.…And nobody would know that Louis Sullivan ever existed. Anyway, the wreckers didn't see him and all kinds of stuff fell on him and buried him. And he died.

I had met with Richard Nickel, oh, maybe a month or so before it happened. The way he talked, oh God, about beauty and past and history and how we must hang on to some things and continuity and all that stuff, I guess you would have to say he was crazy.[37]

In Terkel's anecdote, Nickel becomes a caricature. In his quest for a good story, Terkel neglects to describe the serious work Nickel did to assure that parts of the Stock Exchange Building would end up at the Art Institute. Nickel was also a tireless advocate—and archivist. He in fact made many visits to the Stock Exchange and knew exactly what he was doing, even on his last trip there. He was always aware of the risks that he was taking, as is documented in his letters to friends and colleagues. The Richard Nickel Archive is a testament to his monumental preservation efforts.

IV Information or Object?

7 What Are We *Really* Trying to Preserve: The Original or the Copy?

There is a complex relationship between preservation and copying. We "preserve" some objects by making "preservation copies." In other cases we preserve copies because either they are the only versions of items that exist, or else the copies have their own value. There is some evidence to suggest that in the digital realm, our relationship to copies and copying may be evolving. This chapter will consider some of these issues, with the exception of copyright, which is beyond the scope of this book—although copyright reform will certainly be a monumental challenge. Copyright influences how we produce, access, and use information, but here I am interested in considering the essence of copying itself.

Hillel Schwartz writes that "copying makes us who we are.... Cultures cohere in the faithful transmission of rituals and rules of conduct. To copy cell for cell, word for word, image for image, is to make the known world our own."[1] In a series of essays about twins, doppelgangers, self-portraits, decoys, and other forms of replication, Schwartz makes the case that copies are not inauthentic and are part of our cultural inheritance. He has a point. We learn by copying: infants mimic the sounds of their parents, school children copy sentences from the blackboard, art students copy the paintings and sculptures of well-known artists, forgers study how to copy signatures. Copying can be kinetic, even reflexive. It is an educational tool as well as a means of deceit. We are all shaped by it.

Copying is also the means by which we transmit information and knowledge; it "transforms the One into the Many."[2] The earliest transmission of knowledge was oral; people memorized stories that were passed down from generation to generation. Sometimes the person who memorized the stories copied or mimicked the ways in which storytellers conveyed them. Once writing was developed, oral culture could be transmitted through writing (though oral transmission continued). Texts were copied from manuscript to manuscript, and later from manuscript to printed editions. (Editions were also copied.) There were professional copyists as well as other trades and professions whose work was deeply connected to copying, the aim of which was to preserve and transmit: archivists, scribes (also known as *scriveners*), notaries, tachygraphers, university stationers, secretaries, and court stenographers. Stenography transfers spoken word into printed words; the transfer of the oral to the written continues to the present. Drawings were copied, too; one such copying device, the pantograph, was created in 1603 for draftsmen. It is still in use for many applications.[3]

Other technologies have evolved to simplify copying. In the late eighteenth century, the invention of the copy press made it possible to reproduce holographs.[4] Carbon paper was invented in 1801.[5] In 1839, Henry Fox Talbot, one of the pioneers of photography, created negatives by using paper soaked in silver chloride; the images he copied were fixed with salt. This made it possible for him to copy his photographs. Soon after, microphotographic processes would be developed that could be used to copy texts as well; by the mid-twentieth century microfilms of newspapers were being created.[6] Many kinds of mimeograph and photocopy machines were developed at the end of the nineteenth century and throughout the twentieth. Xerography, developed in the 1930s, was the most pervasive copying device of the second half of the twentieth century.[7] Today scanners make it possible to create digital copies from paper or film. Digital texts and images can also be copied onto a variety of platforms. For example, cutting and pasting digital information from the Web to a Word document is common practice.

It may be convenient for us to be able to make copies easily, but how do we assure the accuracy and veracity of a digital document that has been copied? (See "It Takes a [Virtual] Village," chapter 8.) And the source document itself could be constantly changing. By the time this book is published, figure 7.1 will no longer be accurate: the Wikipedia entry will

---

Prior to the invention of the printing press, the only way to obtain a copy of a book was to copy it out by hand (see scrivener). Throughout the Middle Ages, monks copied entire texts as a way of dissemination and preserving literary, philosophical, and religious texts.

Figure 7.1
Cut-and-paste fragment from the Wikipedia entry on "Copying," in the original Arial 10.5 typeface

probably have been revised at least several times. There is a lack of fixity with Web content.

Finally, we can now copy onto the cloud. Cloud computing uses a network of remote servers hosted on the Internet to store, manage, and process data. This approach is an alternative to saving information on a PC, local server, or network. The implications for this new approach to copying are not yet known—as perhaps is implied by the word *cloud*. Clouds float in the sky, but they may obscure what we can see. And they may break up and disappear, just as so many companies that support cloud content will do. What would John Ruskin make of cloud content? He saw clouds as holding meaning for him, as I explain in chapter 9. He would want us to save in original form as much as we can of our cultural heritage (figure 7.2).

Copying may be an essential aspect of communication, but there is an inherent tension between fixity and evanescence. In many instances, copying causes errors and information is often transmitted incorrectly or lost. Errors may never be caught, yet copies survive and can themselves be reproduced ad infinitum. This necessarily raises issues of accuracy, reliability, authenticity, integrity, and security. While the threat of information loss has always existed, there is quite a bit of anxiety about it today in part because we can lose a great deal—and quickly. (For example, who hasn't

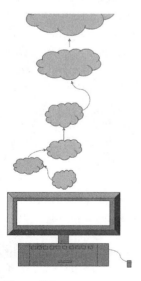

Figure 7.2
Cloud computing. Graphic by Vanessa Reyes, used with permission

lost data in transferring something from one version of a mobile phone to another?)

However, authenticity and reliability have always been considered with regard to copies of documents. Courts of law must often assess the authenticity of documents presented as evidence. In the United States, the Federal Rules of Evidence spell out the requirements that copies must meet to be admissible in court. For example:

When the only concern is with getting the words or other contents before the court with accuracy and precision, then a counterpart serves equally as well as the original, if the counterpart is the product of a method which insures accuracy and genuineness. By definition in Rule 1001(4), *supra*, a "duplicate" possesses this character.[8]

A copy of a public record is admissible as follows:

The proponent may use a copy to prove the content of an official record—or of a document that was recorded or filed in a public office as authorized by law—if these conditions are met: the record or document is otherwise admissible; and the copy is certified as correct in accordance with Rule 902(4) or is testified to be correct by a witness who has compared it with the original. If no such copy can be obtained by reasonable diligence, then the proponent may use other evidence to prove the content.[9]

There are other venues in which the reliability of documents is important: for example, in determining whether something is a fake or forgery, and in research. The methods of determining/establishing reliability and authority differ according to the context. In a court of law, for example, a document may need to be sealed or notarized. To determine whether a work of art is a forgery, one may need to study the provenance of an item. One can also draw evidence from the object itself. Paper, ink, and pigments can be tested for age and for chronological authenticity. This is the domain of experts such as bibliographers, conservators, and scientists.

Digital documents can also be authenticated using techniques such as digital watermarking and other means of content protection such as steganography, which conceals information within other text or content.[10] Important texts, such as legal documents, are more likely to be saved than are, for example, informal communications. Because this is so, resources will be expended to assure digital fixity by every possible means: watermarking, migration, emulation, redundancy, and so on.

At the same time that we are anxious about losing information, our tolerance for digital errors seems to be increasing. Here is one example. For a decade, beginning in 2005, Google engaged in a massive digitization project—usually referred to as the Google Books Library Project or Google Books Project—with large libraries in the United States and Europe; some twenty million books were scanned. The Internet Archive has also engaged in large-scale digitization; to date it has scanned 2.4 million books; approximately one thousand per day.[11] Paul Conway studied the error rate in digital surrogates created by Google and deposited in the HathiTrust Digital Library. At the time of the study (2011) he found that errors occurred in about 1.25 million volumes, or roughly 12 percent of the HathiTrust corpus.[12] A 2007 study by Paul Duguid presented some of the inherent difficulties in scanning books that he concluded contributed to errors.[13] He examined some early editions and copies of *The Life and Opinions of Tristram Shandy, Gentleman,* by Laurence Sterne, a book well known for its unconventional typography and layout. The book contains a black page, a marbled page, blank chapters, arrangements of asterisks on pages, and so on. The copies that Duguid examined were missing some of these features, as well as volume numbers, and other important information.

An early digital library was Project Gutenberg, which was started in 1971 by Michael S. Hart, who, with his volunteers, entered texts by hand in plain text using ASCII. A plain-text version of the bibliographically complex *Tristram Shandy* could not even capture all of the text, let alone the special typographical elements. For example, the two lines of Greek text on the title page were transcribed as

(two lines in Greek).

Duguid assumed that the more innovative Google Books Library Project versions would be more sophisticated than the Project Gutenberg version. While scanning is faster than hand keying the text, and results in a visual version, thus seemingly more accurate, the scanned copies simply introduced new problems. Yet, according to Duguid, many people insist that "innovation should supersede inheritance."[14] Implicit in this statement is that the new is an improvement over the old. But one might equally hold that the new causes us to lose some of the quality of the old.

It is clear that many technologies have given us tools for copying, but they have also given us tools for deception—hence the anxiety about the accuracy and authenticity of copies. Likewise, some copying technologies engender anxiety based on their impermanence. Paper and high-quality microfilm have long lives; digital technology—still primarily driven by commercial interests—has impermanence built in. Planned obsolescence keeps the creators of hardware, software, and new devices profitable. And yet we are dependent on this technological infrastructure for copying, preserving, and transmitting information. Obsolescence keeps us anxious when our responsibility is to maintain long-term access to authentic, authoritative, and reliable documents.

From the beginning of the Google Books Library Project to the present, I have polled the library and information science students in my preservation courses about acceptable error rates in digitization projects. I ask them one question: "If you were managing a large-scale digitization project in your library, what error rate would you be comfortable with?" A decade ago, the answer ranged from 2 to 6 percent (occasionally someone would say that 0 percent was acceptable). The percentage has gradually increased. When I asked the question again in 2016, the average was 10–12 percent, with one student stating that she would be comfortable with a 15 percent error rate. While this is not a methodological approach, and the number of students that I have surveyed is not large—350 students in my preservation classes over 11 years—it may indicate, for error acceptance, changing attitudes among budding librarians and archivists. As one student recently said in class, "What choice do we have?" Another student wondered whether the net gain of having so many items available online might not offset the loss. These people now manage or will soon be managing digitization projects. Has the error-tolerance rate gone up? If so, what are the long-term implications for future scholarship? And what does this tell us about people's attitudes about preservation? If the digitizers of the future are comfortable with a 12–15 percent error rate, how much of our cultural heritage will we lose? I could not identify any studies that have examined how attitudes are changing so it is not possible to know how (or if) such changes will impact future digital projects.

What does this rising tolerance for errors, which I experience with my own students, tell us of our standards for scholarship? One scholar, needing a particular page in a monograph (cited in a text she was working with), sought out that volume in Google Books, only to find that the very page that she needed had not been copied in the scanning project. A 12 percent rate of error (Conway's calculation) may just as well have been a 75 percent rate. If this were medicine, in which lives are in jeopardy because of errors, the percentage would be much closer to zero. Errors in copying create problems where they do not exist in the originals—a strong argument for keeping originals at hand, even when they have been copied.

One other thing can be maddening. Examining the library literature on preservation microfilming from the 1980s to the early 2000s, I have observed that the declared tolerance for filming and scanning errors has remained low. For example, the 2001 specs for preservation microfilming at one institution stipulated that there not be any filming errors. A 2006 article by Karen Coyle about scanning pointed out that even if a top-of-the-line scanner was 99.9 percent accurate, you would still average one error per page.[15] Yet Conway and others have shown that error rates have been higher than the error-rate tolerance recommended in the professional literature. Conway writes that "to preserve the products of large-scale digitization is a decision to preserve imperfection.... For after all, preserving imperfection is an acknowledgement of the deep relationship between the material nature of our print culture and the equally certain physical aspects of our digital world."[16]

Conway's study demonstrates that whatever the tolerance rate for errors is, the actual error rate is much higher. My interactions with students indicate that today's information professionals are comfortable with higher error rates than professionals were a generation ago.

Does that mean that there is a tolerance for imperfection in copies? That is not an easy question to answer. More likely is that we expect less of copies. Most art museums now allow visitors to take pictures of objects on display so long as flashes are not used. The pictures that one takes with a smartphone camera can in no way match the high-quality reproductions that a museum can create. But for most people that is just fine—they are trying to capture the moment rather than venerate or formally record the work of art.[17] This suggests that copies are indeed their own genre, as Schwartz proposes. If that is the case, then there is no reason for a copy to be compared to an original, except in a court of law, an auction house, or other particular contexts.

Schwartz tells us that "practical distinctions between the unique and the multiple have historically been entrusted to theologians, notaries, connoisseurs, and curators."[18] In the digital realm, copying may have taken on its own identity. We may further ask whether copying adds value to the original because it is worthy of being copied. If that is the case, what is the value of the copy itself?

The value of the copy is also determined by the culture in which an object is created or exists. For some, copying may have no legitimacy, while for others the original may not be important. The context for copies—and their preservation—is essential to the stewardship of cultural heritage.

Anything that is not copied risks disappearing. (Again, in some cultures that is accepted and expected.) Schwartz and Conway remind us that copying is ultimately imperfect, but since time immemorial it has been a consistent approach to preservation. Today copying is particularly useful for low-value materials, or any items that are published or created on poor-quality paper. We must fully understand, however, what we are losing by making copies. While creating surrogates is considered by many to be one aspect of preservation, it will never be an ideal solution.

This is a book about our cultural heritage—its monumentality, and the monumental effort we must take to preserve it. Our heritage exists in and is manifested by innumerable "objects" in the analog world, and also, increasingly, "things" digital (converted to digital form or born that way). These objects and things reveal our cultures in many ways. To preserve what they stand for, and to allow them to continue to reveal and represent our culture, we try to maintain the originals. For some things, like books and some prints, for example, we may rely on multiple copies. But for other things—paintings, manuscripts, sculptures, buildings, bridges, certain textiles—there will be only single, unique exemplars. If anything is threatened—or is made uncertain for any reason (as I have discussed throughout this book)—the cultural information it contains may be saved in the form of a copy. But as I have shown, copies are imperfect; copies may themselves be evanescent; and copies may be unreliable or inaccurate or inauthentic for several reasons.

There are untold numbers of these "things" in the world—things revelatory of our culture. What we make copies of in the name of preservation is tremendously problematic. What do we copy? Who will decide and what medium will be used? Is that medium reliable, tamperproof, and affordable? Who will pay for it? Do we rely on analog methods of copying or can we trust digital methods? Are digital copies really *preservation* copies? Given the profit motives of those driving digital technology (with their constant updating of hardware and software), what kind of reliability and longevity (and long-term access) can we expect from digital copies made today?

These are but a few of the myriad questions we could (and must) ask if we look at copying as a form of preservation. As I said earlier, making surrogates will never be an ideal solution. The task of doing so is beyond monumental. But we must face it nonetheless or we risk losing key elements of our culture before we realize they are gone.

8   It Takes a (Virtual) Village: Some Thoughts on Digital Preservation

[Arthur] Sullivan responded to the Iranian protests out of an engaged commitment to human freedom. When the Iranian regime began its repression, he tirelessly documented the horror. Such bitter memories, *now digitally preserved*, give longevity to hope.
—Lee Siegel, "Twitter Can't Save You" (2001)[1]

Love your Kindle? Thank MIT. Media Lab associate professor Joseph Jacobson is a cofounder of E Ink Corp, the company that produces the highly readable black-and-white screens found on many electronic books, including Amazon.com's popular eBook reader. ... E Ink is making life miserable for the printed word, *while preserving the written word in the digital age.*
—"MIT 150," *Boston Globe* (2001)[2]

Digital preservation refers to strategies used to keep born digital and analog information reformatted into digital form both accessible and usable. In 2007 the American Library Association (ALA) developed short, medium, and long definitions of it. The medium definition states that "digital preservation combines policies, strategies and actions to ensure access to reformatted and born digital content regardless of the challenges of media failure and technological change. The goal of digital preservation is the accurate rendering of authenticated content over time."[3]

The preservation of digital content cannot always be guaranteed, even when strategies for its preservation are in place. Thus the two quotations at the head of this chapter are overly confident. Siegel cannot assume that adequate procedures were in place to preserve the digital information that he is referencing, nor is E Ink preserving the written word. This is an ongoing challenge that the preservation community faces; there are few easy explanations that can be made to the public about how to preserve digital objects and information—or how complex such efforts may sometimes need to be. We now live in an information mash-up—preservation may include information that exists in fragments, such as photos posted on the Web or a partial string of emails, or information that has been repurposed.[4] This chapter considers some new approaches to and attitudes about preserving digital heritage.

Neither the ALA medium definition nor the longer one includes the social, cultural, or ethical dimensions of preserving digital content. They also do not account for the ways in which preserved information may be digitally repurposed or consider how digital information may be altered by hackers—revealing how such materials are not always authorial or authoritative (or in modern parlance, authentic). Digital preservation includes not just the methods used to preserve the information—such as emulation or migration—but all of the issues that must be considered to ensure ongoing access to objects, records, and materials. Now that ALA's definitions are a decade old, it is time to consider some of the broader dimensions of digital preservation—and consider a new definition.

In chapter 1, I presented a schema that suggested new directions in preservation. These new approaches are particularly appropriate for digital materials, which often require a network of people to achieve, capture, and save the information: creators, institutions, users, and specialists. This chapter provides three examples of networked/distributed preservation: social media, computer games, and digital art. It also touches on similarities to the preservation of intangible art, the 3D representation of endangered monuments, and the role of ethics in preserving digital content.

Social Networking

As I suggest throughout this book, our behaviors have an impact on how we approach preservation. Social behaviors stem from the culture in which we live (or were raised), our values, societal norms, economic and market forces, and other external factors. To give but one example, the United States is sometimes characterized as a disposable culture with respect to consumer goods. We tend to throw away our possessions fairly soon after we acquire them because they are often inexpensive, or plentiful, or else they quickly become obsolete. Corporations bank on this; they lure us with new and newly desirable products. There is always something "better" just on the horizon. And that horizon is not necessarily far away. A new model of a computer printer I purchased ran out of ink just six weeks after I acquired it. I went to the store to buy a new cartridge and was told that the printer was now obsolete. Digital technology evolves at a remarkably fast pace, creating one of the great impediments to the preservation of materials in digital form.

Compare the life-span of consumer goods today to those of the past, in which longevity was often lauded as a desirable feature. Our parents and grandparents purchased goods with the idea that they would last for decades. My first personal computer lasted for over ten years! Today, we refer to PCs, laptops, or tablets that are three years old as "antiquated" or even "obsolete." We treat objects differently if their shelf-life is three years rather than twenty. Thus, as I suggested in chapter 7, in discussing error rates in digitization, our expectations about longevity and our tolerance for losing data are still evolving. Or perhaps I should say *de*volving, as our expectations about the longevity of objects are decreasing. (Maybe someday we will lease everything.) Expectations of product longevity are no longer rooted in our culture and this is clearly reflected in modern advertising. No longer do we read or hear that a product will give us years of dependable service. The emphasis now is on affordability, convenience, availability, or modernity.

Social network sites (SNSs) are one expression of social networking. SNSs have been defined by danah m. boyd and Nicole B. Ellison as

[W]eb-based services that allow individuals to (1) construct a public or semi-public profile within a bounded system, (2) articulate a list of others with whom they share a connection, and (3) view and traverse their list of connections and those made by others within the system. The nature and nomenclature of these connections may vary from site to site.[5]

The authors also use the term *networking*, which emphasizes relationships. Here I refer to networking, SNSs, and networks. Networking is a social activity that exists with and without technology, though it may result in the formation of networks. It is easier to preserve social network sites than it is to preserve the activity of networking, though networking can certainly take place online on SNSs, and thus some networking activities can be preserved if there are people to keep them going. I use the term *network* in a more general sense to describe an interconnected group of people.

Craig Blaha's study on preserving Facebook records has shown that people who use such social networking sites expect their personal information to persist "indefinitely," which they characterize as less than ten years. Blaha found that the Facebook subscribers in his sample differentiated between "indefinitely" and "forever."[6] (This distinction between "forever" and "indefinitely" is startling. Ostensibly they are almost perfect synonyms. But to young people with no sense of longevity, the words denote distinctly different lengths of time. This may be generational, but for my purposes here it reveals key attitudes about preservation.) His respondents indicated that they may alter their sharing behavior based on their preservation expectations. For example, Facebook subscribers may disclose less information if they believe that their records will exist for a long time. At the same time Blaha's study participants believed that Facebook users created "records of historic value" that should be preserved but they also recognized that the Facebook corporation would probably preserve records only for business purposes, not text of a more personal nature. Finally, the author found that it is not likely that Facebook users will preserve their own records.[7] Thus "archivists should not count on individual subscribers to preserve the corpus of Facebook records despite [Richard] Cox's claims about the 'citizen archivist' in reference to Web-based personal archives. As [Catherine] Marshall puts it, 'the more comfortable people are with computers and the more they can do with them, the more [information] they stand to lose.'"[8]

It is clear from Blaha's study that a great many people using social media think little about the longevity of materials they are creating on SNSs. In fact, they seem to assume that their texts will "always be there" for them. They may have little notion of preservation, and they are not aware of the issue of personal information management (PIM) that is increasingly common in our cultural heritage institutions, especially libraries and archives. To reiterate what Marshall says, people creating content may wish it to be available for the long term, but they do nothing themselves to ensure its longevity. And "the long term" is vague: as long as they [think they] need the information? As long as they are alive? As long as forever? Digital preservation is an elusive concept as much as it is an uncertain phenomenon.

Further, Blaha's study points to ways in which market forces and personal expectations will affect the preservation of social media. More studies are needed to understand how people plan to preserve their personal information. Will they back up their own information? depend on cloud storage systems? or use other strategies? Or will they do nothing, expect others to preserve their materials for them, and hope for the best?

Gaming

While people may not expect their own information to be preserved on social networking sites, there is a strong impetus to preserve some forms of digital media. Gaming is one example. There is tech nostalgia for early games. For example, gamers may seek to preserve the slow speed and screen flicker that characterized the older games, or physical components such as joysticks, which are appreciated as artifacts of the past.[9] Gaming has now been around for long enough that there are many forms to preserve: mainframe-based and video arcade games; consoles; and online games such as first-person, massively multiplayer online role-playing games, and massively multiplayer online games (called in the players' own parlance MMORPGs, or MMOs). With all of these, it is sometimes possible to preserve player cultures, which will make it possible for game historians to understand a crucial component of some forms of gaming. But such an understanding will exist as memories only unless hardware and original (and evolving) software are saved.

Derek Murphy describes an initiative to preserve the last few hours of the virtual world game *EA-Land* at which a "See You Soon Party" was hosted before the publisher "pulled the plug" on July 31, 2008. A student from Stanford University's How They Got Game project of the Stanford Humanities Lab logged into the program and recorded the event.[10] This ethnographic approach complements technical approaches to game preservation, such as emulation, which mimics the original hardware and software. Murphy suggests other approaches to preserving the social aspects of gaming such as ethnographic writing, video documentation, and the crowdsourcing of contributions from the player community itself. Using game players as "citizen archivists," as Murphy proposes, is a grassroots approach to preserving an important part of contemporary culture. This is akin to PIM in that the *users*, not the games' manufacturers, are expected to save the games—and to help others to preserve the experience. Researchers such as Jerome McDonough have used metadata to capture some social information.[11]

There are several ways to preserve. For example, Richard Bartle provides a historical view of game preservation in his article "Archaeology Versus Anthropology: What Can Truly Be Preserved" by drawing parallels between early works such as the Magna Carta and Homer's *Odyssey* and computer games. If the original physical form perishes, copies can be preserved. "Strictly speaking, game historians don't *have* to play the games they study, any more than the historians of Ancient Rome have to participate in gladiatorial combat; however, if we preserve the games for them anyway, at least they have the option of trying them out or not."[12] A key point for Bartle, and the crux of his position, is that game worlds are places and people live there. Players are an intrinsic part of the historical context of games. Therefore, we must archive *now* what people will need to know in the future so that they can understand the games.

But what we archive now is tricky. Do we save original games and the packages they came in? Original fliers and instructions, and other literature that came with the games? The original tapes and disks? For fully online games, do we save all of their manifestations—all of their iterations and editions? Where do these reside? Who has access to them? How can we find these people? How can we convince them (if they exist and we can locate them) to give us these manifestations of the games they enjoyed enough to save? Where will the fiscal resources to preserve the games come from? Who will do the work to preserve them and for how long? And there is another component of preservation we see in gaming: the experiences people had in playing the games in the first place. Can experiences be preserved? The questions I raise here are reminiscent of—and conjure up the same sense of possible futility of—the efforts we must expend in *all* preservation, of buildings, art, history, and all other manifestations of culture.

Digital Art

Digital art is a central part of contemporary art production, though its preservation may present new challenges to museums. For one thing, digital art is dynamic not static, so digital objects can't be "stabilized" in the way that analog media could be. Sabine Himmelsbach likens digital art to performance art because new versions can arise with each viewing.[13] While Himmelsbach identifies the oft-considered challenges to preserving digital objects—technological obsolescence, proprietary software that is no longer supported by the manufacturer, and so on—she also identifies cultural practices that impact preservation. For example, digital culture consists of practices, not just objects. Technology, media, and social relations are part of a process that calls for a networked approach to preservation. That means that institutions have to adapt their procedures to fit the performative aspects of digital art. This includes working closely with the creators of the objects, changing exhibition and documentation practices, and creating a network of experts in digital art conservation. (Most museums don't have such expertise in house.) Himmelsbach's own institution, the House of Electronic Arts, in Basel, Switzerland, focuses on four aspects of preservation: storage, migration, emulation, and reinterpretation.[14]

There are some ways for us to preserve some digital art. Reinterpretation is an important aspect of its preservation. It refers to such practices as replacing an old piece of software with a new one, switching platforms, housing new software in an old "dumb" terminal, so as to appear to re-create the original look of a digital creation. (As with the preservation of old games, it is often desirable to preserve the look and feel of digital art pieces.) Reinterpretation may be the only possible strategy for preserving a digital work. It has antecedents in the analog world: for example, rebinding an old book, or replacing old building materials with new ones. Of course there are purists in all realms of preservation who feel that reinterpretation compromises the authenticity of the original. Such purists can trace their roots to John Ruskin (see chapter 9). But the stakes are higher now; benign neglect is an option in the analog world—not so in the digital world where time is often of the essence in preservation. Further, a reinterpreted iteration of a piece is better than the memory of one that is completely gone.

Or is it? As with performance or environmental art, some digital artists may intentionally create works that will decay; they embrace the inherent temporality of their art. The transience lends to the performative and dynamic aspects of digital works. However, it also means that institutions must understand an artist's intent and possibly make decisions that are appropriate to that intent. As with any performative art, it means that institutions have the responsibility to document and archive the work. In cases in which the artist wants the work to survive, an active artist/museum partnership will need to exist for as long as possible. But if the artist is indifferent about the longevity of a piece, or is outright against the work's survival, what is the museum to do?

In short, reinterpretation requires libraries, archives, and museums to take a network-based approach to preservation. That network may include—but is not limited to—artists, manufacturers, audiences, and outside experts, all with their own ideas of what to save and how to do it.

The opposite of reinterpretation might be a *wabi-sabi* approach to digital art. This Japanese concept gives aesthetic value to the deterioration that an object accrues through aging and use. Wabi-sabi embraces transience and imperfection. Objects that are imperfect, impermanent, or incomplete have value, as in the example of old vinyl recordings. Such an embrace would need to incorporate the concept of loss. The problem would be to control how much loss could accrue. Such an approach could, ironically, require great care to ensure that a digital object didn't disappear. With the example of vinyl records, do we save mint copies, ones with a little background noise, ones with lots of scratches, and so on? Where do we draw the line? How many copies can we save? And in what condition should the copies be to merit our preserving them?

Richard Rinehart and Jon Ippolito express some of the same ideas as Himmelsbach's in their *Re-Collection: Art, New Media, and Social Memory*.[15] In a chapter called "Death by Institution," they deftly describe ways in which museums, libraries, and archives can move away from their historical focus on storage as a preservation strategy to deal with "rapidly mutating media technologies."[16] As does Himmelsbach, they identify migration, emulation, and reinterpretation as the key elements of digital preservation. However, they keenly note that museums tend to collect rare or unique physical objects and tie the value of works to their rarity. Thus museums might be less likely to invest in redundant and distributed data storage, which is critical in guarding against the loss of digital objects. Rinehart and Ippolito warn that sometimes "the lowliest agents on the institutional totem pole"[17] wield the most power when it comes to standard procedures for storage and other policies—and they may enforce the most conservative interpretation of their jobs. Of course, standard professional practices sometimes seem to stifle creative thinking. Cultural heritage institutions must find ways to support innovative thinking among staff at all levels.

As I have observed elsewhere in this book, the notion of *preservation* can be slippery, protean, fuzzy, evolving, and unstable. It is a matter of inter-pretation. So for any work of art—especially digital art—interpretation can lead to many manifestations of a preserved piece.

UNESCO and Digital Heritage

In October 2003, UNESCO adopted the Charter on the Preservation of Digital Heritage[18] to affirm the role of digital preservation in their Mem-ory of the World Programme. The charter includes recommendations for

digitization projects, partnerships, advocacy, publication of guidelines, training and research, and public access to digital heritage. More recently, in
2012, UNESCO convened an international conference, "The Memory of
the World in the Digital Age: Digitization and Preservation," in Vancouver,
Canada.[19] A rationale for the conference was that there was still a strong need
for information professionals to raise awareness of the risks of loss of digital
heritage. Concern was heightened due to the rapid increase of personal, governmental, and commercial information that had been created since 2003.
Further, more work was needed in the areas of digitization policy guidelines
and legal frameworks that could facilitate long-term preservation. In March
2016, UNESCO and IFLA published guidelines that addressed many of these
concerns.[20]

Yola de Lusenet has identified parallels between digital heritage and
intangible heritage. She references the Convention for the Safeguarding of
the Intangible Cultural Heritage as well as the Charter on the Preservation of
Digital Heritage, both published in 2003.[21] Chief among the parallels is that
"discussions on intangible heritage have, in recent years, seriously influenced the thinking on preservation and heritage and have dislodged them
from their solid base of materiality."[22] Intangible cultural heritage refers to
the practices, expressions, activities, and knowledge of communities and
groups. As Article 2 of the Convention says, intangible cultural heritage
"is constantly recreated by communities and groups in response to their
environment, their interaction with nature and their history, and provides
them with a sense of identity and continuity, thus promoting respect for
cultural diversity and human creativity."[23]

De Lusenet, describing the discussions leading up to the charter and convention, observes that some people advocated for including "cyberculture"
in the Convention for the Safeguarding of the Intangible Cultural Heritage;
the proposal was not accepted. For de Lusenet and others at the time, digital
media was best relegated to *documenting* intangible heritage. The Charter and
Convention and de Lusenet's 2007 article predate smartphones, SNSs—such
as Facebook—and other innovations that have facilitated social networking,
and, by extension, online communities. A reading of the Convention for the
Safeguarding of the Intangible Cultural Heritage and the Charter on the Preservation of Digital Heritage today makes the parallels between intangible and
digital communities seem inevitable. Bartle's statement from earlier in this
chapter, that "game worlds are places and people live there," only reinforces

the parallels. However, rather than include digital communities in the Convention on Intangible Culture, I think that in the future we might wish to bring the concept of intangible communities into the Charter on the Preservation of Digital Heritage. Clearly, "the preservation of digital heritage" is one of those simple-sounding but monumentally challenging concepts. We have already spoken of the impossibly large realm of *heritage*: all of the physical, social, psychological, fiscal, historical, and cultural implications it has. And we all know of the tremendously rapidly growing world of digital content. It should be clear to all that we can preserve only a tiny fraction of what exists digitally now and what will be created in the future. The UNESCO charter must deal with procedures of capture, prioritization, funding, weeding, longevity, and so forth. And we must be content with the idea that only a small amount of digital content can be saved—and possibly not even "preserved" in the classical sense in which long-term access is possible.

Financial Sustainability

Digital preservation requires a financial infrastructure that is far more sophisticated than what was once required for analog preservation—namely an institutional budget that could be augmented by external sources such as local and federal grants and private philanthropy. Those sources are still important. However, sustainable, long-term funding for the large-scale infrastructure that is necessary for digital preservation to thrive is required to ensure that digital information will survive. Several years ago the Blue Ribbon Task Force on Sustainable Digital Preservation and Access issued a comprehensive report that outlined strategies for sustainable economic models.[24] The Task Force focused on four categories of information that are of long-term interest to preserve: scholarly discourse, research data, commercially owned cultural content, and collectively produced Web content. The challenges to preservation are long-term time horizons, diffused stakeholders, weak incentives, and lack of clarity about the responsibilities among stakeholders. Absent national policies for preserving digital heritage, there are numerous collaborative projects around the world, including public-private partnerships. Future activities will need to include modifying copyright laws to foster digital preservation as well as the creation of incentives for private parties and corporations to preserve their own assets. The key point is that this kind of preservation is expensive. To achieve it to the best of our abilities, we must secure long-term funding from as many places as we can: corporate, institutional, governmental, and private. Given the amount of data being created now, we must act quickly.

3D Re-creations and Reconstructions as a Preservation Strategy

As described in chapter 4 on Syria, many new technologies are being used to document objects that are endangered by natural or human-made forces. These include the use of cameras on drones, laser scanning, digital modeling, and 3D representations. Such technologies make it possible to document vulnerable heritage sites, and may at least help us record the existence and look of an object. In some cases, it may be the only record. As de Lusenet described earlier, digital media *can* be used to document heritage. A decade ago the tragedy engulfing culture and people in Syria probably couldn't have been imagined. But the point is this: new sophisticated digital tools, in a world where civil wars are rife, are useful for cultural heritage documentation. Such technologies will no longer play subsidiary roles in heritage preservation, for they allow us to re-create originals.

But such re-creations will never be *authentic*. They cannot substitute for the original object. They do, however, serve a purpose. Digital scanning has evolved to such a level that authentic representations can be made of books and manuscripts and they may suffice for some purposes. In the case of war-torn countries like Syria, the technologies allow us to create a historical record of the places that have been heavily damaged or destroyed. However, no copy can substitute for the *feel* and experience of the original.

Ethics and Values

Ethics usually refers to principles regarding conduct while *morals* refers to personal relationships and beliefs. *Values* are socially determined. Samuel Jones and John Holden observe that "in choosing what things we conserve, and how to conserve them, we simultaneously reflect and create social value."[25] As has been described throughout this book, these concepts sometimes bump against one another, as for example with the Bamiyan Buddhas and whether/how to save them (chapters 1 and 10). Ethical standards and guidelines are abundant at the international level (UNESCO and international professional associations such as IFLA, ICA, and ICOM), the national level, through regional associations, and in local institutions. Thus, digital preservation practices are guided by a host of international, national, and institutional ethical standards. Yet there may be some practices that do not neatly fit into ethical standards.

Peter Johan Lor and Johannes J. Britz tackle some complex issues in their essay "An Ethical Perspective on Political-Economic Issues in the Long-Term Preservation of Digital Heritage." They present two hypothetical cases,[26] to challenge the notion that the preservation of digital heritage is inherently "good." (The authors sent the second case study to thirty-three librarians in developing questions for their feedback.) The first case deals with the harvesting and preservation by a wealthy country of the cultural heritage of a less affluent one. The second case concerns a project that a wealthy country initiated to digitize the cultural heritage of a less affluent country. The studies are in a social justice framework that aligns information rights with human rights. The authors further consider who has the rights to preserve digital heritage.

In the first case, a wealthy nation (called "Opulentia") downloads the websites of an opposition party in a poor country (called "Povertopia"). The authors pose ethical questions such as whether it is right to download the sites without permission, whether such harvesting is justified in the name of scholarship, and whether we can make the case that we (the people of Opulentia) are helping the people of Povertopia by preserving a part of their heritage that they may not be able to preserve for themselves.

The second case describes a cooperative digitization program between libraries in Opulentia and Povertopia for which Opulentia provides the resources for digitization and storage, and maintains control of the archives; it charges a modest fee for access, except for students and scholars in Povertopia who can access it for free. Responses to the second case study from the librarians who were contacted is telling. In favor of the program it set forth were those librarians who are concerned that materials in their countries are disappearing at an alarming rate. At least one responded that he or she would accept the proposal because her [his] country is poor and lacks the resources to digitize. Weighing in against such a partnership were respondents who raised copyright issues, felt that partnerships should be equal, or thought that the terms of limited access were unfair.

While these cases are hypothetical, it is likely that similar projects are actually taking place. These could raise ethical and moral concerns. A primary concern is asymmetrical power issues among nations. These have certainly occurred in the analog world where it has been easy for wealthy people or nations to take artifacts from less wealthy nations. But there are other issues as well. One has to do with values and perceptions. Will countries that are "digital have nots" live in the "digital dark ages" if they cannot afford to use the digital technologies that "digital haves" access easily? Is it possible to survive in a world in which digital information predominates, without the capacity to engage in such a world? Lor and Britz present a convincing case that we must pay close attention to the ethical dimensions of digital preservation. To do so will enhance social justice generally.

As with many topics considered in this book, ethical issues are multi-dimensional. Cultures clash over many things, and yet, despite their different opinions, they can justify their positions using the "truths" of their religions, their laws, their social systems, and so on. Cultural "norms" vary from people to people. In the preservation field we must be culturally sensitive in our dealings with others—particularly, in the present context, with respect to the handling of digital information. Do we have a right to invoke our own values when we are deciding how to deal with information from other countries, peoples, or cultures—information we wish to preserve (or to let die)? Should our values—even if they clash with those of other cultures—help us to decide to preserve those others' information resources? We need an extensive examination of ethics and values, our own and others', to guide us in our preservation of cultural heritage.

Yet there are already many exciting opportunities for sharing cultural heritage. Abby Smith Rumsey describes the richness and vitality of the digital world when she writes that "now with the Internet, we can continue the conversation [among generations and across the ages] as we digitize materials from the past, but also broaden it across multiple languages and civilizations with different historical experiences and expectations of the future."[27] This chapter has suggested ways in which we might broaden our definition of digital preservation by embracing its rich, social aspects.

The social complexities we have seen in this book—complexities across cultures—make our own tasks of defining and acting on preserving our heritage difficult. We need to look at the digital world in a circumspect way. Institutions cannot assume sole responsibility for preserving digital heritage. Instead, there must be a global and virtual "village" to assure such preservation. Sabine Himmelsbach's description of a "network of care" for digital art applies equally to all digital content. (Aptly, figure 8.1 illustrates that "now" might be of short duration.)[28]

Figure 8.1
"How Long Is Now" (Berlin). Photograph by Michèle V. Cloonan, 2016

A New Definition?

We can define *digital preservation* so as to include its rich social dimensions. I propose a new, fuller understanding along these lines: Digital preservation includes personal and collaborative efforts to maintain content in a form that is accessible for as long as it is needed. Such efforts will be undertaken by the owners of the content, the creators of it, heritage professionals, and users. The goal of preservation is the accurate rendering of authenticated content over time that takes into account the many uses and modes of digital heritage and communication with a sensitivity to all those creating, saving, and using it, and all others impacted by it. Digital preservation is dynamic and its strategies must constantly evolve. Therefore, we must consider not only the preservation of content, but also the preservation of the context in which it was created and in the world in which it will be used.

V The Greening of Preservation

9 Sustainable Preservation

Thunderstorm; pitch dark, with no *blackness*,—but deep, high, *filthiness* of lurid, yet not sublimely lurid, smoke-cloud; dense manufacturing mist; fearful squalls of shivery wind. ... I never saw such a dirty, weak, foul storm. It cleared suddenly after raining all afternoon, at half-past eight to nine, into pure, natural weather,—low rain-clouds on quite clear, green, wet hills.
—John Ruskin, "The Storm-Cloud of the Nineteenth Century," 1884 lecture[1]

There are many links between the environmental movement and the preservation of cultural heritage, stemming from our long-standing interest in how we protect the built environment—and nature. What are the origins of these links? What do the fields have in common? What are the points of intersection? What can we learn from the natural world? How does the concept of sustainability play out in the various heritage fields? The aim of this chapter is to compare the conservation, preservation, restoration, and stewardship of the natural environment to that of the built environment and human-made objects and records. The commonalities and differences among the fields of environmental studies; historic preservation; and library, archives, and museum preservation have been little studied. Yet while these are distinct fields, all three have examined, for example, what it means to try to bring something back to a previously known state, whether it is a forest, a painting, a statue, or a document. And in all three fields, there have been tensions between control and stewardship. Can museum goers touch the objects? Can ranchers allow their livestock to graze on protected lands? Who owns (and therefore controls) heritage?

Writings related to preservation concerns date back in the West to antiquity. The need to preserve records is expressed in the Bible, for example in Jeremiah: "Thus said the Lord of Hosts, the God of Israel: 'Take these documents, this deed of purchase, the sealed text and the open one, and put them into an earthen jar, so that they may last a long time" (32.14); and in Isaiah: "Now, go, write it down on a tablet, and inscribe it in a record, that it may be with them for future days, a witness forever" (32.14).[2] In Genesis, the relationship of human beings to nature is set out: God commands humanity to "be fertile and increase, fill the earth and master it; and rule the fish of the sea, the birds of the sky, and all the living things that creep on earth" (1:26–28).[3] In 15 B.C.E., Vitruvius advocated a preventive approach to the preservation of the built environment in *On Architecture*,[4] and centuries before that, Plato wrote about man's relationship to nature. Gabriela Roxana Carone makes the case that Plato advocated "for the preservation of species of nonhuman animals, which in many relevant ways are not ranked below the human species."[5] There has been a long-standing interest in how we cohabitate with and preserve the natural world, and how we preserve human-made heritage—perhaps stretching back as far as human communication.

While writings in all of these fields date back centuries (see table 9.1), the nineteenth century is a logical starting point for considering heritage studies. Many things occurred then that directed attention toward preservation. Romanticism in visual art, literature, and music focused on nature. Those who embarked on Grand Tours of Europe wrote extensively about classical ruins. The industrial revolution inspired the Arts and Crafts Movement, which, with its appreciation of the past, advocated for historic preservation. More pragmatically, the nineteenth century paved the way for new professions such as art history, library science, archival science, and, in the twentieth century, historic preservation and the environmental sciences.

Other international events focused attention on heritage. European settlers took land from indigenous peoples in the United States, Canada, Australia, Africa, and elsewhere around the world. The new arrivals became the dominant force depriving indigenous peoples of their land and forcing the native inhabitants to assimilate to the newly predominant and largely white Christian culture or to move to undesirable land. Cultural genocide often ensued, as the indigenous peoples were stripped not only of their land, but also of their languages, customs, and ways of life. An early action on the part of the U.S. government was the Antiquities Act of 1906 (16 U.S.C. 431–433) to protect Native American archeological sites on federal lands.[6] However, many other violations against Native Americans—and

Table 9.1

A Selection of Preservation and Conservation Activities in the Nineteenth and Twentieth Centuries. Table by Michèle V. Cloonan

---

Cultural Heritage

19th century

Preservation

1851–1853: John Ruskin writes *Stones of Venice*.

1853: Ann Pamela Cunningham and others found the Mount Vernon Ladies' Association to preserve President George Washington's home near Alexandria, Virginia.

1877: William Morris and others write the Society for the Protection of Ancient Buildings (SPAB) British manifesto on the preservation of buildings.

1895: The National Trust is founded in Great Britain to protect and preserve historic places.

1897: Ancient Shrines and Temples Preservation Law is enacted in Japan.

1899: The first of two peace conferences to consider the protection of cultural property is held in The Hague. These conferences underpin the 1954 Hague Convention for the Protection of Cultural Property in the Event of Armed Conflict.

20th century

1906: U.S. Congress passes the Antiquities Act of 1906 to protect Native American archeological sites on federal lands.

1910: William Sumner Appleton and others found the Society for the Preservation of New England Antiquities (SPNEA), now called Historic New England.

1931: The Athens Charter on the Restoration of Historic Buildings is enacted.

1949: The National Trust for Historic Preservation receives its charter from the U.S. Congress.

1964: The International Charter for the Conservation and Restoration of Monuments and Sites is enacted, better known as the Venice Charter for the Conservation and Restoration of Monuments and Sites.

1966: The National Historic Preservation Act is signed into law, launching a national system in the United States.

1976: The Tax Reform Act in the United States provides some tax incentives for rehabilitating historic properties.

1979: The Australian National Committee of the International Council on Monuments and Sites adopts the Charter for the Conservation of Places of Cultural Significance, better known as the Burra Charter.

1990: Native American Graves Protection and Repatriation Act (NAGPRA) is passed by the U.S. Congress.

1990: The U.S. National Trust for Historic Preservation adopts the Charleston Principles, eight guidelines for community conservation.

1994: The Nara Document on Authenticity originated in a conference held in Nara, Japan, sponsored by ICOMOS.

Table 9.1 (continued)

Environmental Heritage

19th century

1864: George Perkins Marsh publishes *Man and Nature.*

1865: Commons Preservation Society (now the Open Spaces Society) is founded.

1872: U.S. Congress establishes Yellowstone Park, the country's first national park, which signals the U.S. government's acceptance of responsibility for national heritage conservation.

1884: Ruskin's The *Storm-Cloud of the Nineteenth Century* (crossover with historic preservation) is published.

20th century

1916: The National Park Service is established in the United States.

1919: The Historic Sites, Places of Scenic Beauty, and Natural Monuments Preservation Law is introduced in Japan; it includes flora and fauna.

1945: The United Nations Educational, Scientific and Cultural Organization (UNESCO) is founded, an early commitment to world cultural and natural heritage.

1962: Rachel Carson's *Silent Spring* is published.

1963: The Clean Air Act passed in the United States (amended many times since).

1969: The U.S. National Environmental Policy Act (NEPA), a broad national framework for protecting the environment, is passed, enacted in 1970.

1970: The Environmental Protection Agency (EPA) is established in the United States.

1972: UNESCO enacts the Convention Concerning the Protection of World Cultural and Natural Heritage (also called the World Heritage Convention).

1992: United Nations Framework Convention on Climate Change, known as the "Earth Summit," is held.

1992: Operational Guidelines for the Implementation of the World Heritage Convention introduces the term "cultural landscape" to refer to combined works of nature and people.

2016: Statement on UNESCO's Environmental and Social Policies is issued.

indigenous peoples in other countries—have continued and remedies for these actions are still ongoing (see chapter 3), not only from governments, but from businesses as well.[7] The United Nations estimates that there are more than 370 million indigenous people in 70 countries.[8] Preservation issues will abound for the foreseeable future.

Against this complex nineteenth-century backdrop is a person who lived for much of the century—and wrote on preservation. John Ruskin (1819–1900) was a keen observer, writing about architectural preservation and also about clouds (figure 9.1). He spent much time looking at clouds

Figure 9.1
Portrait of John Ruskin, attributed to J. Lindsay Barry. Courtesy of the Ruskin Foundation, Ruskin Library, Lancaster University

and considering them from his perspective as an artist, art historian, and moral commentator. He drew extensively on his diary entries—which span decades—in his lectures and publications. While his observations about clouds were often metaphoric, he writes in the preface to the published version of his two "Storm-Cloud" lectures (1884) that his aim is to "bring to your notice a series of cloud phenomena, which, so far as I can weigh existing evidence, are peculiar to our own times; yet which have not hitherto received any special notice or description from Meteorologists."[9] He refers to what he calls the "plague-cloud,"[10] describing it this way:

It looks partly as if it were made of poisonous smoke; very possibly it may be: there are at least two hundred furnace chimneys in a square of two miles on every side of me. But mere smoke would not blow to and fro in that wild way. It looks more to me as if it were made of dead men's souls—such of them as are not gone yet where they have to go, and may be flitting hither and thither, doubting, themselves, of the fittest place for them.[11]

This passage encapsulates the various ways in which we can read these lectures. Ruskin's words could reveal his observation of what is actually happening in the real world: human-made pollution is changing the physical world. But was he revealing real "poisonous smoke" or was he speaking metaphorically and morally? Is his writing really descriptive of the literal environment? Undoubtedly he could see air pollution—given the "two hundred furnace chimneys" around him—but his point has to do with the cause, and that might be a sign, he believed, of what was happening to humanity. (And even a casual observer could see the black grime on London's buildings, clearly related to the black clouds and, as we know, a nasty product of the Industrial Revolution.)

Are Ruskin's writings moral and metaphoric, or do they presage environmental writing? A portion of the February 4, 1884, lecture is included in the *Norton Anthology of English Literature*.[12] The editor writes, "Some newspapers complained that the lecture seemed merely to blame air pollution on the devil; however, what Ruskin was blaming was the devil of industrialism, the source of the 'Manchester devil's darkness' and of the 'dense manufacturing mist.'"[13] Brian J. Day observes that the essay is "environmentally conscious" but that it is not an environmental piece per se. For Ruskin, pollution has moral implications: it signals an alienation from God *and* nature.[14] Day characterizes Ruskin's essays as exhibiting "moral ecology." Ruskin sees pollution as "a problem that is essentially spiritual in origin."[15] For Ruskin, nature as perceived by humans has a "moral index."[16]

Ruskin compares his observations about the sky to those of scientists:

And now I come to the most important sign of the plague-wind and the plague-cloud: that in bringing on their peculiar darkness, they *blanch* the sun instead of reddening it. And here I must note briefly to you the uselessness of observation by instruments, or machines, instead of eyes. In the first year when I had begun to notice the specialty of the plague-wind, I went of course to the Oxford observatory to consult its registrars. They have their anemometer always on the twirl, and can tell you the force, or at least the pace, of a gale, by day or night. But the anemometer can only record for you how often it has been driven round, not at all whether it went round *steadily* or it went round *trembling*. And on that point depends the entire question whether it is a plague breeze or a healthy one: and what's the use of telling you whether the wind's strong or not, when it can't tell you whether it's a strong medicine, or a strong poison?[17]

Ruskin is critical of scientists as he believes that his powers of acute observation and insight, honed over many decades, are superior to scientific instruments of observation. Further, he is convinced that science must exist within a moral framework: environmental pollution stems from moral pollution. Ruskin did not discover a new type of cloud, except, perhaps, as a metaphor for the ills of late nineteenth-century England. He was aware of environmental changes resulting from the Industrial Revolution in Britain. That he closes his essay on a note of "rectitude and piety" shows us the many levels on which Ruskin was writing.

Ruskin's observations can be seen as something of a "prequel" to environmental ethics, "the discipline in philosophy that studies the moral relationship of human beings to, and also the value and moral status of, the environment and its non-human contents."[18] Ethical issues continue to inform our consideration of the environment. (See, for example, figure 9.2.)

Figure 9.2
"Clouds-over-Skyline," gelatin silver print, by Vik Muniz. Permission of Art © Vik Muniz/Licensed by VAGA, New York

And one suspects that today Ruskin could write an essay on climate change, once again informed by his own observations.

Ruskin's writings on storm clouds are appropriate in a chapter that illustrates parallels between the preservation of the built and natural environments. Ruskin wrote on both subjects, drawing on his study of nature and of architecture, and, as I said, infusing a moral perspective. His "Lamp of Memory," in *The Seven Lamps of Architecture*, continues to have a strong influence on historic preservation today because he advocated a minimalist approach to caring for buildings that is attuned to current practice. If we care for a building in its *existing* form, the integrity of the building will be maintained. Ruskin argues strongly against restoration:

Neither by the public, nor by those who have care of public monuments, is the true meaning of the word *restoration* understood. ... Do not let us deceive ourselves in this important matter; it is *impossible,* as impossible as to raise the dead, to restore anything that has ever been great or beautiful in architecture. That which I have above insisted upon as the life of the whole, that spirit which is given only by the hand and eye of the workman, can never be recalled. Another spirit may be given by another time, and it is then a new building.[19]

More famously, in the same essay, he wrote:

Do not let us talk then of restoration. The thing is a Lie from beginning to end. ... But, it is said, there may come a necessity for restoration! Granted. Look the necessity full in the face, and understand it on its own terms. It is a necessity for destruction. Accept it as such, pull the building down, throw its stones into neglected corners, make ballast of them, or mortar if you will; but do it honestly, and do not set up a Lie in their place. And look that necessity in the face before it comes, and you may prevent it.[20]

For Ruskin, architecture represented cultural memory; "ruining" a building through restoration impacted culture as a whole. Preservation represented man's inheritance. Similarly, he believed that all new buildings must be designed and constructed in such a way as to guarantee permanence for future generations. There is a stewardship component of Ruskin's vision; he believed that if buildings were properly cared for, they wouldn't need to be restored. "Take proper care of your monuments, and you will not need to restore them," he wrote, adding, "Watch an old building with anxious care; guard it as best you may, and at *any* cost from every influence of dilapidation."[21]

Ruskin also refers to the destructive "restoration" of the abbey of St. Ouen, which was rebuilt, but not according to the original design. He is referring to his French contemporary, Eugène-Emmanuel Viollet-le-Duc

(1814–1879), though he does not mention him by name, who practiced "interpretative restoration." This approach did not seek to return a building to its original state—which Ruskin and others declared to be impossible anyway—but to save the building. Viollet-le-Duc was an interventionist.

Even today, preservationists debate about the boundaries of restoration. Ruskin describes an ideal state in which minimal treatment is always possible. But of course intervention is sometimes necessary. Historic preservation recognizes that different approaches need to be taken in different circumstances. And this is also true in the conservation of objects as well. The more culturally significant an object may be, the more attention it may get from conservators—but too much treatment can also lead to deterioration. Many well-known works have been over restored, for example the Sistine Chapel (more on this to follow).

Ruskin bemoans the state of architecture in his own time; some buildings were not necessarily worth preserving. How might Ruskin approach the Leaning Tower of Pisa? Would he have advocated letting it collapse? The tower was constructed from 1173 to 1372, but began leaning during construction due to an inadequate foundation, and ground that was too soft on one side to adequately support the structure's weight.[22] The tilt was recently partially corrected; but not too much.[23] The goal was to stabilize the tower to prevent it from collapse, while preserving the tilt. After all, the tilt is what makes the tower a tourist destination. Is this an example of necessary intervention? Ruskin had opposed the restoration of another structure in Pisa, the Santa Maria della Spina (ca. 1230), which was dismantled and rebuilt on higher ground in 1871 to protect it from potential flooding from the Arno River.[24]

For Ruskin there was a moral dimension as well: only moral men could create moral architecture. Today, there is still a "moral imperative to preserve" our heritage though this imperative is not placed so much on individual character as it is on collective actions.[25] Such an imperative is expressed in the chapter titles of Rachel Carson's *Silent Spring*, for example: "Elixirs of Death" and "Beyond the Dreams of the Borgias." Her work drew attention to the ways in which the widespread use of poisons to control insects was changing the balance of nature, and her language conveys that theme powerfully. For instance, in her section on "Elixirs of Death," on insecticides, she writes: "They have immense power not merely to poison but to enter into the most vital processes of the body and change them in sinister and often deadly ways."[26] Other authors have weighed in on the environment in strong terms. Mahatma Gandhi observed that "Earth provides enough to satisfy every man's needs, but not every man's greed."[27] Filmmaker Rob Stewart has written that "[t]here is simply no issue more important. Conservation is the preservation of human life on earth, and that, above all else, is worth fighting for."[28] While Ruskin focused, in the natural world, on storm clouds, writers in the past century have looked at the many threats to preservation: pesticides, commercial greed, water pollution, and so on. Their writings echo Ruskin's.

Ruskin's views influenced William Morris, one of the founders in 1877 of the Society for the Protection of Ancient Buildings (SPAB), an organization in Britain that is still active. Melanie L. Jackson, in her master's thesis on Viollet-le-Duc, Ruskin, and Morris, concludes that modern preservation organizations have built on all of their principles, although the principles of Ruskin and Morris have dominated. While there are still negative opinions about restoration, elements of it are integrated into contemporary preservation activities. Jackson illustrates how particular preservation organizations (e.g., SPAB, the United States National Trust for Historic Preservation, and the International Council for Monuments and Sites) have incorporated aspects of these views.[29] Today, we recognize a middle ground that accepts that some restoration may be necessary to preserve a building. More than that, we recognize that there must be balance in our approach to preservation. Therefore, charters, conventions, and laws will continue to evolve; the Venice Charter and the Burra Charter are but two examples of this. Through preservation management, professionals seek to balance protection and intervention.

Ruskin wrote about clouds and buildings, but he didn't integrate these writings into a generalized view of preservation. It is probably fair to conclude that no one in the nineteenth century did. Today we understand that poor air quality affects not only buildings, but also books, documents, and objects as well (not to mention what it does to those living in and breathing bad air).[30]

Much research has been devoted over the past century to determining "ideal" environmental standards for storing and displaying artifacts. Managing the environment is integral to good library and museum practice. However, views about what constitutes ideal storage facilities have evolved from absolute temperature and humidity standards to optimal preservation environments. This has occurred for three reasons: the high costs of maintaining certain conditions, the need for heritage institutions to reduce their carbon footprints, and the understanding—born of research—that there are greater acceptable temperature and humidity ranges than we once believed. It is more sustainable for an institution to use less energy.

The groundwork for the integration of cultural and natural heritage was laid by UNESCO when it was founded in 1945. The 1972 World Heritage Convention (WHC) linked the concepts of nature preservation and the preservation of cultural sites. In 1992, a revision to the WHC introduced the term "cultural landscape" to identify the combined works of nature and man. Ken Taylor, Archer St Clair, and Nora T. Mitchell note that this resulted in "bridging the traditional nature-culture dichotomy in the heritage conservation field."[31] The authors enumerate issues that have grown in importance since 1992, including progressive abandonment of the countryside in large areas of the world, climate change, and technological change. Additionally, increasing attention is now paid to the role of landscape in our sense of place and identity.

Scholars now consider that the built environment should be a part of environmental ethics. Roger J. H. King believes that "[a]n environmentally responsible culture should be one in which citizens take responsibility for the domesticated environments in which they live, as well as for their effects on wild nature."[32] By creating a built world responsibly we respect nature. One way to do that is to design sustainable buildings. Another way, as Paolo Soleri pioneered, is to build using the natural materials from the very site of the building.

Environmental sustainability concerns, and includes, the entire built environment. *Sustainability* is a term used by heritage professionals to refer to responsible stewardship. It can refer to the built or the natural world. In that sense, there is an obvious overlap between cultural heritage institutions and the environmental movement. Another example of environmental sustainability has to do with disaster preparedness. Most institutions engage in some level of preparedness and prevention. Such programs include monitoring buildings, evaluating collections, training staff, procuring disaster-recovery supplies, working cooperatively with fire departments and local and federal agencies, and so on. Disaster preparedness efforts now also integrate weather patterns into planning.

In "Stewarding the Future," David Lowenthal points out that sustainability implies a commitment to manage cultural and natural resources into an indefinite future, yet there is no general agreement as to how far ahead that might be. Is it for a generation or two, or many millennia, or even further out?[33] Whatever the length of time may be, stewardship takes the long view. The word derives from the Old English *stigweard* "to keep" or to "watch out for."[34] Stewardship originally referred to property stewards, but today it is used broadly to embody the management of all resources.

Lowenthal adds that there are many reasons for a "futurist stance." One is ethical: it is only fair that future generations inherit a world that has not been denuded or spoiled. Another is familial: we want our children to live in a world as fruitful, and at least as clean, as our own. But there are also pragmatic issues: intergenerational equity promotes social stability. Stewardship is inherently optimistic and is a corrective to the more pessimistic view that the world is likely to succumb to bioterror, for example,[35] or that it is seriously deteriorating under human influence. Stewardship implies that we can maintain our planet; that is, that the planet will not be in worse shape for future generations.

Jennifer Welchman explores the concept of stewardship from the perspective of virtue ethics.[36] She posits that benevolence and loyalty are crucial for environmental stewardship and that people may be motivated to be stewards of the environment. Such motivations may serve "human desires for happiness, peace, security, beauty, and so forth. In other words, we would want to know what (if anything) motivates people to function effectively as voluntary *stewards* of the natural world."[37] Yet for people to see stewardship as an appropriate role for them, they must learn about environmental processes. Specifically, people need to understand how current human management of the natural world may affect the quality of life of future generations. Equally important is an understanding of how nature has shaped human history and traditions.

These are some of the obvious relationships between the built and natural worlds. However, there are other similarities as well. Janna Thompson has used cultural heritage as a lens through which to view the environment.[38] She has observed that preserving cultural heritage and preserving nature have often been thought to have nothing to do with one another. Yet, many of the reasons that we have for preserving nature have to do with human associations. She presents several examples: a tree in a community that is associated with the early settlers and thus has become a landmark; a notable scenic landscape that is under threat of commercial development; a wilderness area that is associated with national heritage. In short, nature can be valued because of its human associations. Thompson reflects that heritage can include objects, traditions, customs, and environments—natural and human-made.

The terms that we use may reflect our disciplinary approaches. In the heritage fields, *preservation, conservation,* and *restoration* have specific meanings. *Preservation* is the umbrella term for everything we do to maintain cultural heritage, including copying and reformatting. It implies stewardship. *Conservation* is the treatment of individual items and collections, though it also refers to other strategies for collection care. *Restoration* may refer either to locating that which has been lost from the original and reintegrating it— as in replacing lost film footage into a restored film—or, more generally, efforts that are made to bring something back to a known earlier state. For example, a restorer might try to make an object appear as it once looked. In the book and museum worlds *restoration* implies taking a defective object and fixing it in such a way as to hide its flaws—presenting it as unblemished. Of course, as Ruskin pointed out, this is inherently impossible.

These terms are used in the environmental context as well. The U.S. National Park Service defines two of them this way:

Conservation and preservation are closely linked and may indeed seem to mean the same thing. Both terms involve a degree of protection, but how that protection is carried out is the key difference. Conservation is generally associated with the protection of natural resources, while preservation is associated with the protection of buildings, objects, and landscapes. Put simply, conservation seeks the *proper use of nature,* while preservation seeks *protection of nature from use.*[39]

During the environmental movement of the early twentieth century, two opposing factions emerged: conservationists and preservationists. Conservationists sought to regulate human use while preservationists sought to eliminate human impact altogether. Aldo Leopold, often called the father of ecology, called for wilderness protection and an enduring land ethic. Wilderness preservation is fundamental to the idea of deep ecology—the philosophy that recognizes an inherent worth of all living beings, regardless of their instrumental utility to human needs.[40]

David Baron expressed the differences this way:

For a century, environmentalism has divided itself into warring camps: conservationists versus preservationists. ... The struggle pits those who would meddle with nature against those who would leave it be. ... The only sensible way forward lies in a melding of the two philosophies. If nature has grown artificial, then restoring wilderness requires human intervention. We must manage nature in order to leave it alone.[41]

Restoration is the renewal, through active intervention, of degraded, damaged, or destroyed landscapes. Marcus Hall begins his book with an analogy between the restoration of damaged landscapes and damaged art, using the treatment of Michelangelo's *Last Judgment* in the Sistine Chapel for the comparison. This most recent "restoration"—there have been several since the paintings were finished during the Renaissance—resulted in controversy about the methods used as well with the results.[42] Hall cites sources that detail the controversies,[43] some having to do with the overpainting that took place (after Michelangelo's death) to cover up the genitals that he painted into the nude bodies, and the removal of the grime that, over the centuries, dulled the artist's original colors.

But the crux of the matter for Hall is whether anything can be brought back to its original state. "In what is sometimes referred to as the 'Sistine Chapel Debate,'" he writes, "restorers of natural systems wonder to what extent they can or should be returning these systems to their original forms. Can tall-grass prairies be reestablished on farmland that has been growing corn and alfalfa for more than a century?" Hall gives many other examples, and points out that there can never be a "restoration project that gives perfect results." In the case of a work of art, we can almost never know what it looked like when it was new, and what the effects of the environment have been on it. The Sistine Chapel presents a complex combination of preservation challenges. Its frescoes suffered from hundreds of years of damage to the chapel's walls. It has been exposed to and endured smoke from incense used in church services; air pollution and dust that streamed into the chapel from open windows; millions of tourists passing through; and many restorations of varying quality. Given these challenges, some art historians and conservators believe that the most recent treatment of the Sistine Chapel was successful and restored vitality to the frescoes.[44]

A neo-Ruskinian approach, writes Salvador Muñoz Viñas, would be to do nothing at all; the painting would never be cleaned and the original varnishes would be maintained, because such intervention would remove historical evidence, such as "a follicle or a bit of skin from the artist."[45] Writing about the cleaning of Michelangelo's sculpture *David*, Muñoz Viñas quotes James Beck's assertion that "the study of … DNA could in the future give unexpected information."[46] Beck's position could be seen as a truly sustainable approach to conservation. However, as Muñoz Viñas points out, an evidential approach to conservation would serve historians, archeologists, and scientists, but would not be satisfactory to the public. Of course, such a debate is moot for Michelangelo; his works have undergone restoration for centuries and his *David* was most recently cleaned in 2003–2004.[47]

David A. Scott takes these ideas further by exploring the concept of authenticity in conservation, drawing on the Athens and Venice Charters, the Nara Document on Authenticity 1992, and UNESCO World Heritage documents as well as the writings of eminent art historians and conservators. For Scott, authenticity rests on three principle foundations: the material, the historical, and the conceptual. He shows that there is no unified concept of authenticity and thus there is no universally accepted approach to conservation, "no synthesis of information between art connoisseurship and scientific connoisseurship."[48]

Whether we are considering the preservation of natural systems or artworks, we must also consider whether it is better to restore something or allow it to deteriorate (art) or decay (nature) further. As complex an undertaking as restoring the Sistine Chapel has been, Marcus Hall points out that while works of art are created by artists over days or possibly years, "landscapes result from countless animate or inanimate forces acting over centuries or millennia" ... [and] for severely degraded sites like an open-pit mine or a toxic dumping site, land managers may aim to restore only those elements necessary for sustaining rudimentary plant or animal life."[49]

Hall characterizes the early conservation and preservation debate as the "utilitarian camps, that is ... those who were motivated by aesthetics and those motivated largely by pragmatics."[50] Some conservationists cared primarily about the land's beauty while others cared about its productivity. Gifford Pinchot (1865–1946) was a conservationist who looked for ways to put resources to work, while John Muir (1838–1914) sought to protect scenic wonders such as Yosemite. However, as I have pointed out, in the quotation from the National Park Service, as heated as the debates were, conservation and preservation imply stewardship of the land, even if the means of doing so are different.

A component of conservation that is shared by the built environment and land management is the profit motive. In chapter 6 we saw the obstacles that Richard Nickel faced when trying to save historic buildings from commercial developers who could profit by tearing the structures down. Similarly, land conservation (and preservation) may also face challenges from business and tourism. Sometimes divergent interests come together: a restored train station is turned into profitable shops and restaurants and a neighborhood is renewed, or forests can be profitably managed for continuous cropping.

Hall suggests that conservation can also be divided into preservation and restoration, which puts the focus on management *methods* rather than management *motives*. Muir advocated leaving land untouched, while Pinchot called for replanting forests that had been cut down. Pinchot sought to bring back former abundance (restoration), though Hall describes him as a utilitarian restorationist rather than merely an aesthetic one. "'Restoration,' a common word used in the early literature of land management, must be written back into the history of the conservation movement," he wrote.[51]

George Perkins Marsh (1801–1882) compared American and European approaches to conservation, particularly those in the United States and Italy, where he was a foreign minister. The aims of his book, he wrote, were "to indicate the character and...the extent of the changes produced by human action in the physical conditions of the globe we inhabit; to point out the dangers of imprudence and the necessity of caution in all operations which...interfere with the spontaneous arrangements of the organic or inorganic world."[52] He recognized that people were causing problems to the environment. Much as Ruskin had identified changes in the sky, Marsh was studying the land. Logging and grazing were affecting the habitat. For example, the removal of vegetation in watersheds led to flash floods. Marsh had observed this in the Alps of northwestern Italy; later in the century, the same thing happened in the Rocky Mountains of central Utah. Both situations created the need for enormous vegetation restoration projects.

What is the proper relationship between humans and nature? Is nature something to be conquered and manipulated? Georges-Louis Leclerc, Comte de Buffon (1707–1788) believed that humans could improve the land; in fact, enlightened humans brought good changes to the land.[53] This view was popular in Italy, as well.[54] As mentioned at the beginning of this chapter, that relationship has been discussed for many centuries. By the nineteenth century, Marsh broke new ground by studying the relationship in detail not only in the United States and Italy, but in other countries as well. He described the threat of deforestation and asserted that land can be conserved only if people properly manage it. He concludes that

we are never justified in assuming a force to be insignificant because its measure is unknown, or even because no physical effect can now be traced to it as its origin. The collection of phenomena must precede the analysis of them, and every new fact, illustrative of the action and reaction between humanity and the material world around it, is another step toward the determination of the great question, whether man is of nature or above her.[55]

Hall returns to the "Sistine Chapel Debate":

Ultimately, it may seem that those who set out to bring back the Sistine Chapel are restoring culture, while those who bring back wildland are restoring nature. Yet there is clearly much more culture and nature than one realizes in both chapel and wildland—whether one restores like Buffon or Marsh or [Aldo] Leopold. Whether as gardeners or repairers or naturalizers, restorationists depend on both culture and nature, but they make very different assumptions about the role of each in the restorative process.[56]

Hall makes one last parallel to the Sistine Chapel with respect to calls for the Glen Canyon Dam across the Colorado River to be dismantled, contemplating the outcome given the thick layers of gray silt all around the canyon. "As with Michelangelo's masterpiece, one wonders if this canyon could ever be brought back—or whether it should be restored at all. Unlike restoration of the Sistine Chapel, one wonders how far nature, itself, could be relied on to wash away the silt and erase the defects, to re-awaken the canyon wren's call and replant the cottonwoods, penstemon, and scarlet gilia.[57]

In the 1960s, the Sierra Club's David Brower battled those who wanted to build a dam in the Grand Canyon. Proponents asserted that tourists in boats would have better access to the canyon. Brower retorted with four full-page ads in the *New York Times* and other periodicals. His question "Should we also flood the Sistine Chapel so tourists can get nearer to the ceiling?" is now well known among environmentalists.[58]

The Sistine Chapel restoration resonates with environmentalists. Is it because it is a "monumental" work of art, on par with universally appreciated landmarks, such as the Grand Canyon or Mount Everest? Or is it because any restoration effort of something so universally recognized is bound to be controversial? Or is the Sistine Chapel simply a symbol of the impossibility of truly restoring anything?

* * *

In restoring land, to what state can we ever bring it back? We can trace the history of land management as far back as we have records. But we do not know much about the relationship between indigenous people and their land before new settlers took control of it. Bill Gammage has studied how the Australian Aborigines managed their lands. In 1788, Aborigines and

Europeans made first contact after the Europeans arrived at Sydney Cove. And "1788" is shorthand for the practices of Aboriginal people at the time they first came into contact with whites. Gammage uses the term *1788 fire* to refer to the deliberate use of fire as a land-management strategy.[59] His book includes early European descriptions of the landscape; the new settlers could not account for the fact that there seemed to be so few trees, for example. In fact, it wasn't until the twentieth century that researchers began to understand how the Aboriginal people managed their land—and the planting and placement of trees. Gammage himself has played an important role in studying it, and he learned that the Aborigines used fire and the life cycles of native plants to make sure that there would be ample wildlife and plant-based foods throughout the year. Gammage identifies three rules to 1788 Aboriginal management:

- Ensure that all life flourishes.
- Make plants and animals abundant, convenient, and predictable.
- Think universal, act local.[60]

According to Gammage these rules imposed an ecological discipline on everyone. Land was cultivated without fences, so that farm and wilderness were one. "It made a continent a single estate," he wrote of Australia.[61]

Gammage's research suggests that for the indigenous people in Australia there is an inherent connection between cultural and natural heritage. Those connections deserve to be better understood. At the same time, as Thompson put it, "[t]here is no single historical narrative for society."[62] In countries such as the United States, Canada, and Australia, which have been recently settled by Europeans, the settlers have their own story to tell. However, the European perspective has been the dominant one. A full understanding of Australia's rich and complex history will depend on many readings of it.

The point here is that cultural preservation and even the preservation of the natural world will often require a sensitivity to the long history of that which is being preserved. And this sensitivity must be alert to untouched nature and also nature that has been "managed"—how it has been interacting with people long before recorded history, if that history is recoverable in some way, as through oral traditions or discernible traces of human involvement.

* * *

Connections between the natural and human-made environments are made explicit by the Council of the European Union in a recent report.[63] It is striking how the authors of the report have infused cultural heritage with the language of the environmental movement. In the following passages, I have indicated in italics words and phrases that are used more commonly in the environmental literature than in the cultural heritage sector.

Cultural heritage consists of the *resources* inherited from the past in all forms and aspects—tangible, intangible and digital (born digital and digitized), including monuments, sites, landscapes, skills, practices, knowledge and expressions of human creativity, as well as collections conserved and managed by public and private bodies such as museums, libraries and archives. It originates from the interaction between people and places through time and it is constantly evolving. These *resources* are of great value to society from a cultural, environmental, social and economic point of view and thus their *sustainable management* constitutes a strategic choice for the 21st century.[64]

Cultural heritage as a *non-renewable resource* that is unique, non-replaceable or non-interchangeable is currently confronted with important challenges related to cultural, environmental, social, economic and technological transformations that affect all aspects of contemporary life.[65]

Cultural heritage plays a specific role in achieving the Europe 2020 strategy goals for a *"smart, sustainable and inclusive growth"* because it has social and economic impact and contributes to environmental sustainability.[66]

It seems likely that scholars, policymakers, and citizens will continue to find links between the environmental movement and the preservation of cultural heritage. This chapter has presented some emerging perspectives.

The Monumental Challenge of Preservation is about *monumental preservation*. What can be more monumental for us than the earth itself! And given all the issues raised in this chapter, can we ever reach consensus about what to do with the natural world—let alone the creations of people? The "conversations" about preservation/conservation/restoration of the natural world in several ways do conjure up the conversations about the analog and digital products of human beings. In my view, there are too many phenomena for us to generalize about "best practices"; and there are too many sides to the issues for us to find common ground about what constitutes ideal ways of thinking about preservation. Our worlds (natural and human-made), and people themselves, are so complex that to achieve "sustainable preservation" is beyond monumentality, no matter what realm (human or natural) we are considering. The best we can do is to recognize the issues and the implications they have for all constituencies, and try to find common ground for what our most reasonable avenues of actions could be—the greatest good for the greatest number, or something of that sort.

VI   Enduring, Ephemeral Preservation

10 Preservation: Enduring or Ephemeral?

A decade ago I published "The Paradox of Preservation," an article that considered whether it is possible to maintain objects and collections "*indefinitely or even for a long time.*"[1] I presented five case studies to "challenge our assumptions about what needs to be preserved and how to achieve this."[2] The article proposed that each case represented a paradox, which I defined, following *The Cambridge Dictionary of Philosophy*, as "a seemingly sound piece of reasoning based on seemingly true assumptions that lead to a contradiction."[3] (Today I would expand that definition, as does Merriam-Webster, to include "a person, situation, or action having seemingly contradictory qualities or phases."[4])

The five cases were 1) the discovery of—and damage to—the Nag Hammadi Gnostic scriptures, whose leather covers are the earliest-known surviving codex bindings; 2) preservation of the Auschwitz concentration camp complex; 3) a hypothetical case about the preservation of cartoons of caricatures of Muhammad that were published in the Danish newspaper *Jyllands-Posten* on September 30, 2005; 4) the destruction of the Bamiyan Buddhas in Bamyan, Afghanistan, on March 11, 2001; and 5) the glass-covered ruins of the Hamar Cathedral in Hamar, Norway.

The paradox of the Nag Hammadi codices is that their discovery initially led to their misuse and damage rather than their preservation. They were discovered by farmers and some of the texts were used for kindling; others disappeared. Only ten and a half of the original thirteen volumes wound up in the Coptic Museum—which did not initially have the resources to adequately care for them. Would it have been better for their preservation if they had not been discovered? Scholars maintain that what has been gained in our knowledge of early Gnostic scriptures was worth what was lost and damaged.

The preservation paradox is that had the codices remained buried, if they were later discovered by people who understood the value of what they had found, the volumes might have survived in better condition. However, the important knowledge that was uncovered might not now be ours. Hence, it may have been fortunate that the codices were discovered. Of course, some might say that it is better to let objects remain in the ground, and thus excavation is often subject to debate. Yet every discovery yields new knowledge; in this case we know more about early Christian theology than we did before the discovery.

The issue with the Auschwitz camps in my second case study was whether the structures should be preserved, and if so, in what form. In 1947 the Polish parliament determined that Auschwitz would be "forever preserved as a memorial to the martyrdom of the Polish nation and other peoples."[5] Because they were built to be temporary, some people thought that the buildings should be allowed to fall into ruins, and only the grounds preserved. However, the prevailing sentiment was that it was important to preserve the evidence of the atrocities that were committed at the camps. But then another issue arose: should there be only modest preservation intervention or large-scale reconstruction? The latter approach was taken in recognition that Auschwitz has over the decades become an archive, a museum, and a gathering place, and not just hallowed ground.

The paradox is that it is difficult to preserve something that was intended to perish. In fact, we live in a world full of ephemeral items, many of which we must actively preserve if we are to leave behind us as complete a record of history as is possible. Thus the preservation of all parts of Auschwitz is justifiable, rather than paradoxical. Our libraries, archives, and museums are filled with things their creators saw as ephemeral—but things that add measurably to our knowledge of the culture of the past.

In 2005, twelve professional cartoonists in Denmark each created a drawing titled "Muhammeds ansigt" (The Face of Muhammad), which were then published in the *Jyllands-Posten* newspaper. Not all the sketches were of the prophet; one was of a boy with that name, and one was of a Danish politician. The point of the piece was to challenge censorship. Because Islam proscribes depictions of Muhammed (worship is to God alone), negative responses to the cartoons were swift. Letters to the editor, protest marches, and other nonviolent activities ensued. Ambassadors from Muslim-majority countries wrote to Denmark's prime minister to protest what they felt was an ongoing campaign against Islam and Muslims in Europe. Protests spread around the world, some of which were violent. Danish embassies in some Arab countries were attacked. And there were effective boycotts of Danish consumer goods in Middle East countries. The ramifications from the publication of the cartoons were felt in Denmark for years.

Soon after the event, in early 2006, I contacted six curators of large Islamic collections in American libraries to find out whether they would acquire copies of the newspaper—if they did not subscribe to it—or copies of the cartoons. If they acquired them, would they catalog them? Exhibit them? Or make them available to researchers in other ways?

All six curators whom I polled indicated that they would collect the cartoons, whose importance they recognized. They all emphasized that they would not back away from collecting controversial items. However, while all of them recognized the importance of collecting and cataloging the items, the consensus was that it was probably premature to advertise or exhibit the cartoons given the sensitivities at that time. (Many publications were initially fearful of reproducing them; they are now on the Web.) This response from a curator was characteristic: "Everything is grist for the historian, and, in this case, the culture critics, political scientists, constitutional scholars, and so on, so I would acquire and catalog, as you would any artifacts in print, whether text or image, but not display or advertise."[6]

An issue for libraries and archives immediately following the event was whether it was safe for them to display or publicize the cartoons. Would the curators be opening themselves to attack? One of the respondents observed that

[t]here are many activists within the academic and non-academic world who are trying to extract and ban various collections for many reasons. Libraries have an obligation to preserve "primary sources," including the cartoons that have spurred the riots and the killings. How else can [we] study the violent protests and diplomatic upheaval that ensued [after] the publication of the cartoons?[7]

If the institutions collected but did not display or publicize the cartoons, would their users know to look for them? Would anyone be aware of their existence? Does preservation sometimes necessitate withholding access? If collecting and preserving the cartoons are in order, must the institution be willing to live with the consequences? Even if the items are subsequently destroyed by users? Can an institution safeguard itself? Any decision that librarians or archivists make will undergo the scrutiny of users—who may hold divergent opinions. There are several paradoxes in this case study, the most important of which is that libraries acquire items for *use* so restricting it by not making the cartoons accessible makes no sense. Although it is not unknown for libraries to restrict access to items that are fragile or are in some other way vulnerable to harm, it is not the preferred approach for librarians to take.

Paul Sturges, in "Limits to Freedom of Expression? Considerations Arising from the Danish Cartoons Affair," believes that the starting point for exploring the dimensions of freedom of expression is the United Nations Declaration of Human Rights. He suggests that librarians must "think clearly about the avoidance of harm and offence in balancing human rights" with freedom of expression for library users.[8] He stresses the need for librarians to establish policies for handling controversial materials that strike a balance between disseminating them and making them available for consultation. His views reflect the responses of the professionals whom I interviewed.

The same issue had arisen years earlier with the publication of Salman Rushdie's novel *The Satanic Verses*. After the book was published in 1988, the Ayatollah Ruhollah Khomeini, the former supreme leader of Iran, issued a fatwa on Rushdie in 1989 that remains in place today. There were many things about the book that were offensive to readers, including the title of the book, which references a reputed legend about prophet Mohammed. Killings and bombings resulted over Muslim anger about the novel and Rushdie had to live in hiding for many years. Many libraries removed the book from their shelves, placing their copies in Special Collections departments for safekeeping. This incident exposed a cultural chasm between Western and Muslim values. Sensitivities erupted once again with publication of the Danish cartoons.

Charlie Hebdo, a French satirical newspaper, also published cartoons of Muhammad as well as satires of sharia law. The publications incited violence that culminated in bombings of the *Charlie Hebdo* offices on November 3, 2011, and January 7, 2015. Sturges wrote a follow-up article, "Limits to Freedom of Expression? The Problem of Blasphemy," not long after the January 2015 bombing. In this piece he concludes that freedom of expression is too important to be limited by blasphemy laws; library and information professionals must continue to advocate for freedom of expression in spite of continued attempts to suppress it.[9]

We live in complex times. As I write this, just after the U.S. 2016 presidential election, a number of libraries have reported receiving threats over the past year for simply having "objectionable" materials.[10] This is quite different from challenges that libraries have faced in the past for acquiring specific titles; for perhaps the first time, libraries are facing hate crimes. Among the crimes have been the defacing of "objectionable" books and personal attacks. For example, in one library, a man approached a patron wearing a hijab and tried to remove it. In other instances, swastikas or anti-Muslim messages have been written on walls, windows, or books.[11]

The fourth case study relates to the destruction of the Bamiyan Buddhas. (see figures 1.10 and 1.11). In my article I dubbed it "The Saddest Preservation Story Ever Told," not imagining that as tragic as the destruction of the Buddhas by the Talban was, it would pale in comparison with the even more violent destruction of heritage that has since taken place in Syria by ISIS (see chapter 4). Blowing up the Bamiyan Buddhas was not a sudden assault. Mullah Mohammed Omar had issued an edict against the Buddhas before 2001; in fact, the Taliban were already destroying ancient sculptures.[12] The giant Buddhas from the sixth century, carved out of the sandstone cliffs, were prominent when the Bamyan (also known as Bamiyan) Valley was part of the Silk Road. Buddhists left the area hundreds of years ago, and international advocacy on behalf of the preservation of the sculptures held no sway with the Taliban. After the site was bombed, UNESCO sent a mission there to assess the damage and to protect the remaining stones. In 2009, ICOMOS constructed scaffolding around the niches to stabilize the site.[13] In 2011, a UNESCO Expert Working Group held a meeting in Paris to decide what steps to take. Thirty-nine recommendations emerged, which included leaving the larger niche empty as a monument to the destruction of the Buddhas and rebuilding the smaller of the two Buddhas.[14] After the 2011 meeting, German archeologists from the German branch of ICOMOS, led by Michael Petzet, began work to re-create the smaller Buddha's lower appendages with iron rods and concrete and bricks. UNESCO subsequently suspended that work, pending further study.

But given more recent damage to cultural heritage in the Middle East, should we be trying to restore or even reconstruct what is left of the Buddhas? Won't they just be destroyed again? Some believe that by restoring the site, tourism will increase that would aid the surrounding—and poor—communities, while others believe that money should be spent directly on the communities, not on restoration of the Buddhas.

In a recent article, "Ruin or Rebuild? Conserving Heritage in an Age of Terrorism," Robert Bevan examines contemporary dilemmas concerning the conservation of monuments.[15] While there may be a strong impulse to rebuild, reasons *not* to rebuild must be carefully considered. Whose best interest does rebuilding serve? In the case of the Bamiyan Buddhas, the Taliban was not the only Muslim group opposed to the statues, so reconstruction might be an invitation to further destruction. (The statues were pulverized, making reconstruction with original materials impossible.) This issue is further complicated by the fact that the current local government would like to encourage tourism in the area.

Andrea Bruno, an architectural consultant to UNESCO, has suggested creating a sanctuary at the base of the niche that once held the larger Buddha. The idea was expanded to include a cultural center, with funding from the South Korean government.[16] This may encourage tourism while also respecting the customs and beliefs of the local area.

The Bamiyan Buddhas—and more recently, the war in Syria—remind us that the protection of cultural heritage must focus on people as well as on stones or brick. We need to care not only about heritage but also about the well-being of people. Bevan reports that in UNESCO there is a new emphasis on linking heritage to human rights and humanitarian aid. While this may seem obvious, it is not. We often read about the "world's (shared) heritage," but preservation is also about local politics and communities.

Perhaps the paradox here is one of conflicting perspectives and values; there is no "correct" preservation strategy. We are only a little bit closer to a solution in 2017 than we were in 2001, or 2007, but if the Bamiyan Buddha site becomes a center in which the now-empty niches take on a new cultural meaning, it will stand as a fitting monument for today's world.

The final case study relates to the restoration of the Hamar Cathedral, which was built in the beginning of the thirteenth century in Hamar, Norway. In 1567, during the Northern Seven Years' War, the cathedral was burned down. Later the ruins were used as a quarry and many of the cathedral's stones were used in the construction of other buildings. The ruins further deteriorated over time—and many harsh winters. How might the remains of the cathedral be preserved?

The architectural firm Lund & Slaatto preserved the ruins by encasing them in an enormous protective glass structure—a building to protect what was left of an earlier building. While such a structure could constrict the ruins (see, for example, the glass-encased Kaisersaal at the Sony Center in Berlin described in chapter 11), the glass structure in Hamar is majestic—even monumental. The ruins have been given a new life, not just as a site but as a social center and gathering place. And the ruins should be less vulnerable to further deterioration.

I described the paradox of the Hamar Cathedral this way: "the overall structure—the ruins plus its glass encasement—now constitutes the overall notion of 'cathedral.' The preservation activity yielded a new concept of what the building literally and figuratively stood for."[17]

Today I wonder: is this historic preservation or a historic transformation? The ruins themselves will be less vulnerable to deterioration brought about by harsh weather or vandalism, but they have been recontextualized. This evolution is not natural. The idea was to preserve the cathedral—or what was left of it, including its memory. But the solution has turned the ruins into something different. On the other hand, the new site is graceful and beautiful with a spiritual quality all its own. Perhaps in this case preservation is merely the byproduct of creating something new, and newly meaningful. But the old, original meaning is gone; the structure is no longer a place of worship. It is now a place of commerce and social interaction. We may have preserved some of the original building, but we have not preserved much of the building's original purpose or spirit.

* * *

Ten years on it is hard not to reflect on what has happened in the world since I wrote "The Paradox of Preservation." The fallout from the so-called Arab Spring has left cultural heritage in the Middle East more vulnerable than ever before to archeological terrorism and has led to the destruction of libraries and archives that might hold "objectionable" materials. The Coptic Christians have always been vulnerable to attack—even under the current Egyptian government of President Abdel Fattah el-Sisi, which has been tolerant of them. Recently, a suicide bomber killed twenty-four Christians during Sunday Mass at a Cairo church;[18] other heinous attacks have followed by ISIS. How safe are the Nag Hammadi codices, housed as they are in the Coptic Museum?

The destruction of the Bamiyan Buddhas now seems like a prelude to the many attacks that have taken place in Iraq and Syria since then. Indeed, tensions in the world are increasing, as evidenced by numerous terrorist attacks in Europe as well as in the Middle East. In fact, the rage that many expressed at the destruction of the Buddhas seems mild now in comparison to the destruction we have witnessed in Palmyra and many other Middle East cities.

At the same time, information is increasing at an ever-faster rate, facilitated by new technologies and spread amazingly quickly through social media. As I have suggested throughout this book, we must think about preservation in new ways. One such example: the preservation of records and materials will sometimes need to be information-based *not* material-based. We will need to be prepared to copy information that physically resides in vulnerable parts of the world. While examples of copy-and-rescue projects were given in chapter 4, we will have to prepare for an increase in violence against many manifestations of our culture. With the seemingly unstoppable destruction of monuments that we have witnessed in the last decade (and ISIS has not been alone in its depredations), perhaps the best we can do in our preservation efforts is to make visual and verbal records of important cultural phenomena widely available in digital representations.

Now that we have nearly thirty years of experience digitizing records, and now that we can draw on new technologies such as drones to record information about archeological sites, we are on the verge of a new era of preservation. I have suggested some strategies in chapters 4 and 8. One direct consequence of the recent violence and of our use of new technology for recording information is that cultural heritage communities will need to work more closely with one another than they have in the past. Beyond the LAMs (libraries, archives, and museums) are historic preservationists, archeologists, environmentalists, and other professionals who are concerned with preserving the natural and built world.

* * *

The past decade has led me to think more broadly about paradoxes—indeed, led me to write this book. I now question whether preservation is inherently paradoxical. Perhaps it is more accurate to say that the world is made up of paradoxical situations that require creative preservation strategies, and some of them seem to contradict others. Or, put another way, perhaps it is not preservation that presents paradoxes, but rather the beliefs and practices of people. If it is life itself that is paradoxical, maybe we need to abandon the assumption that it is possible to preserve objects "indefinitely" or "for as long as possible."

Preservation will always mean different things to different cultures. Even within cultures, people will always have contradictory perspectives about what and how to preserve.

At the end of my 2007 article, I stated that "we can preserve some things some of the time; but not everything all of the time, and we cannot operate purely under an old custodial model."[19] A new model that encompasses new technologies and accounts for an increasingly complex world must be developed. One lesson is that everything changes, as with people's beliefs, political parties, religion, and people themselves. We must track those changes and see how our views of preservation must evolve to guide us to the best strategies for preserving all we deem important to save. One of the aims of this book is to raise awareness of these issues and challenge readers to think globally and locally about our responsibilities as preservers of culture. It is a monumental challenge.

11 Epilogue: Berlin as a City of Reconciliation and Preservation

Alle freien Menschen, wo immer sie leben mögen, sind Bürger dieser Stadt West-Berlin, und deshalb bin ich als freier Mann stolz darauf, sagen zu können: "Ich bin ein Berliner!"

All free men, wherever they may live, are citizens of West Berlin, and therefore, as a free man, I take pride in the words "Ich bin ein Berliner!"
—John F. Kennedy, June 23, 1963[1]

A city is a text with many pages, and every page counts. Too many pages are missing from Berlin's urban history.
—Renzo Piano, architect[2]

Contemporary Berlin embodies many of the themes related to monumental preservation that I have explored in this book: how and when to preserve heritage; how to document the past; how to memorialize people and events; how to address the cultural genocide that was inflicted upon millions of people prior to and during World War II; and what gets preserved and what does not.[3] Perhaps, most significant of all, Berliners are showing us that *how* to preserve defies a single solution. Rather, in Berlin as in elsewhere, there are many approaches to preservation.

Berlin has some unique challenges—for example, what it means to preserve the history of a once-divided city. It also faces an issue of growing attention in cultural heritage institutions in countries of former colonial powers: what to do about ancient objects that they acquired and that are now contested. For example, Berlin continues to face ownership claims related to Nefertiti, one of its most famous cultural heritage objects (figure 11.1). (Her image appears on German postcards and stamps.) After the ubiquitous

Figure 11.1
Bust of Queen Nefertiti. Photograph, Rol Agency, France. Courtesy of Europeana

Berlin bears and the Brandenburg Gate, she is the city's most famous icon. The Egyptians maintain that she was taken out of their country under false pretenses, while the Germans assert that she was obtained from the Egyptian government legally (more on this subject to follow).

As rich and nuanced as all of these topics are, there is another aspect of Berlin that makes Nefertiti relevant to this study: Berliners are thoroughly engaged with preservation. As I suggested in chapter 1, with the preservation schema (figure 1.1), the role of the public has become increasingly important to successful projects. As will be seen in Berlin, seemingly every memorial, every teardown, every rebuilding, elicits passionate public response. In 2003 I visited the site of the future "Memorial to the Murdered Jews of Europe" (Holocaust-Denkmal). It was a large, mostly bare area filled with trenches, surrounded by a chainlink fence on which people had hung and taped hundreds of notes with all kinds of opinions and observations about the memorial—positive, negative, indifferent, and irrelevant. Clearly, residents (and probably also visitors) were deeply engaged in this memory project. This shows that residents and tourists throng to construction sites to observe their goings on. There is currently a structure known as the "Humboldt Box" in front of the construction site of the Berliner StadtSchloss (also called Berlin Palace Humboldt Forum, or Humboldtforum) that includes a viewing area (along with exhibitions and tours). According to the website http://humboldt-box.com/, Humboldt Box is "one of the city's most visited attractions." The tradition of creating viewing stations in construction areas probably started during the rebuilding of the Potsdamer Platz in the 1990s. In 2001, Svetlana Boym described Berlin as "a Museum of conceptual art."[4] These vehicles for participation exist alongside the usual forums: the press and other media, casual conversations on the streets, protests, exhibitions, social networking, and so forth. Preservation inheres in this city, and it exists in a realm of *people*, residents and visitors, who, in one way or another, want to be involved. This underscores one of the persistent themes of this book: that preservation cannot take place as an isolated activity. In saving objects, buildings, sites, ideas, and cultures, we must be keenly aware of the human element—the things that impact people's lives.

Berlin was founded nearly eight hundred years ago in 1237, though the past two hundred years are most germane to our preservation "tour" of the city. It was once the capital of Prussia as well as the largest city in the German empire. An Academy of Arts was founded in 1696, followed by an Academy of Sciences in 1700. When the French occupied Germany in 1806, Napoleon Bonaparte's armies made off with many of Berlin's works of art. Beginning around 1810 plans were drawn up for a public museum collection; Berlin gradually established itself as a museum leader. Prussia's first public museum, the Altes Museum, opened in 1830 and housed paintings and sculptures.[5] A decade later, the Neues Museum was created for cultural objects. Museums continued to be founded in the area known as Museumsinsel (Museum Island) until 1930; they showcased some of Berlin's spectacular acquisitions, such as the Pergamon Altar, the Ishtar Gate, and the aforementioned Nefertiti bust, along with many other Egyptian artifacts that were once housed in their own museum and are now in the Neues Museum. Other museums sprang up in the city as well, including the Dahlem Museums complex, which once housed the Nefertiti bust.

Prussia's defeat of France in the Franco-Prussian War of 1870–1871 led to the modernization of Berlin, which was economically and culturally enriched by the outcome of the war.[6] The 1911 entry on Berlin in the *Encyclopaedia Britannica* noted that "in no other [city] has public money been expended with such enlightened discretion." By the end of the nineteenth century, as a cultural hub Berlin could have been favorably compared to London and Paris. In fact, when it came to acquiring foreign art, Berlin often competed with them. Kathryn Gunsch, for example, describes how in 1898 reliefs acquired by the British in Benin City (today in Nigeria), wound up in Berlin.[7]

Two world wars would greatly change Berlin—and the world. (Here our focus is on cultural heritage—not on the wars themselves.) Germany was first a perpetrator and then a victim of art plundering. Under the Nazi regime, Jews were stripped of their German citizenship, and their art collections, libraries, and personal papers were confiscated for Adolf Hitler. At the same time he ordered so-called "degenerate" art and books to be destroyed and their artists and authors to be deported. Later, using a network of Nazi agents, Hitler took artworks of his liking from across Europe for his collection. He was inspired by a 1938 trip to Italy to rebuild Berlin on a grand scale that would reflect his monumental ambitions. Similarly, he wanted to turn Linz, Austria, a city near his birthplace, into a cultural capital with a major art museum—the Führermuseum.

"His" art objects were cataloged and stored in several locations, including air raid shelters, remote castles, and salt mines. Hitler created sketches of his intended museum and cultural complex. Architect Roderich Fick created models for the doomed complex.[8] The postwar disposition of the many thousands of hidden artworks was complex; objects continue to be rediscovered. Some items were looted, some returned to their rightful owners, and some—which rightfully belonged to Germany—were later returned there.

Much of Berlin was bombed, and many artworks that belonged to the museums were destroyed or severely damaged. After World War II Germany was looted by the Allied forces. As with Hitler's collection, many of the Berlin museum objects were never returned to Germany. The plunder was revenge for Germany's wartime activities, but there were also people who stole art for their own personal gain. No matter the cause of the plunder, stolen objects continue to be discovered. Additionally, claims continue to be made by survivors of the families whose artworks and libraries were confiscated by the Nazis.

Some objects that were damaged during the bombings of Berlin survive as memorials to the war. In the Bode Museum there is a small room of damaged statues—burn victims—from the Allied bombings of May 1945. The large wall panel describes the bombings:

The Second World War unleashed by Germany in September 1939 not only led to the damage and destruction of many of Berlin's museum buildings [sic]. It also resulted in an inferno for a large part of the museum's holdings. Between the 5th and 6th of May as well as between the 14th and 18th of May 1945, two devastating fires accompanied by explosions broke out in the main tower of the anti-aircraft bunker in Friedrichshain, which was one of the main evacuation sites for artworks belonging to the Berlin museums. The causes of this catastrophe, which occurred days after the surface air-raid shelter located in a public park had been surrendered to the Soviet armed forces, have never been satisfactorily explained. It resulted, however, in the greatest destruction and damage of art works [sic] in European museum history.

Like many other artworks exhibited in the Bode Museum, the three Italian sculptures on display in this room were recovered from the rubble of the devastated Friedrichshain bunker. They were initially transported to the USSR as looted art and then returned to Museum Island in 1958 along with many other objects belonging to the Berlin museums. More or less damaged, they are expressly presented here in a location whose appearance, with the exception of the wall coverings, corresponds to the time when the museum was built. Exhibited here with all their "wounds," they are cautionary reminders of the effects that irresponsible human events have had on the world's cultural heritage.[9]

One example in the Bode room is this Italian bust (figure 11.2). The Germans continue to maintain a list of *Beutekunst* (looted art)—the thousand or so missing objects that were confiscated by the Americans, Russians, and other Allied forces. The Germans persist in trying to recover their lost objects. The Museumsinsel is testimony to the highs and lows of German history. Itself now historic, in 1999 UNESCO added Museumsinsel to its list of World Heritage Sites. It is also decaying; nearly two billion euros have been spent on the renovation of the Museumsinsel since 1992.[10] (Burned objects are

Figure 11.2
Italian bust at the Bode Museum, damaged in the May 1945 bombings. Image by
Michèle V. Cloonan, April 2016

displayed at other museums in Berlin as well, including the Musikinstru-
menten-Museum, which has a case of burned instruments—with no cap-
tions. The wounded instruments speak for themselves.)

Berlin has many other preservation stories to tell. I have selected a few.

The Reichstag

The Reichstag (the Parliamentary Building) was designed by Paul Wallot,
and built from 1884 to 1894 on land owned by Prussian Count Raczynski,

who refused to sell it in his lifetime. The Reichstag was extensively damaged by a fire in February 1933; Marinus van der Lubbe, a Dutch communist, was executed the following year for allegedly setting the fire. He may or may not have been involved in the arson; he was posthumously pardoned in 2007. Some sources claim that it is more likely that Adolf Hitler had the fire set as a way to persuade parliament to pass the Enabling Act—which greatly increased Hitler's power.[11] The Reichstag was also bombed during the war. It remained damaged and uninhabitable for more than fifty years before it was restored and reconstructed. The German flag was raised on top of the building as part of the German reunification celebrations in 1990; in 1991 the Deutscher Bundestag (German Parliament) voted to return the seat of the German capital to Berlin. Parliament has convened there again since April 1999.

Artists Christo and Jeanne-Claude created *Wrapped Reichstag* in 1995. The project was conceived in 1971, but took many years to bring about. The wrapping material was a silvery polypropylene fabric that captured the contours of the building. The many photos of it on the Web show the large number of spectators who were drawn to it. The Reichstag remained wrapped for two weeks and the materials were subsequently recycled. The wrapping project was a gift to the Germans. (See http://christojeanneclaude .net/projects/wrapped-reichstag for details of the project.)

Jeanne-Claude and Christo's work celebrates impermanence, and their *Wrapped Reichstag* seems now to have been a set piece for an urban stage on which permanence versus impermanence has been the running show for the past thirty years.

Soon after "the wrapping," architect Norman Foster reconstructed and redesigned the Reichstag building (1994–1999). Jeanne-Claude and Christo's installation was a symbolic handing over of the building to Foster. The lengthy process of coming up with a plan for the new parliament that all the German decision makers could agree to is well described in Deyan Sudjic's biography of Foster.[12] The use of glass for the dome was intended to symbolize transparency in government as well as public accessibility. The wrapping project, in its modern (1990s) context, suggested that the building would be handed over—like a wrapped gift—to Foster, and that his work would lead to the building's being handed over to the country. These activities have become part of the permanent record of the Reichstag, preserved

Figure 11.3
The dome of the Reichstag. Image by Michèle V. Cloonan, April 2016

in the documentation of the project. Appropriately, tourists flock to the Reichstag. The dome affords a beautiful 360-degree view of Berlin (figure 11.3).

Brandenburger Tor

The Brandenburg Gate, perhaps the best known of Berlin's monuments, is in the middle of the Pariser Platz, just across from the Reichstag. The gate was designed by Carl G. Langhans and completed in 1789. Perched on top of the gate is the Quadriga, a twenty-foot-high sculpture of a laurel-crowned goddess in a chariot pulled by four magnificent horses. It was designed by the sculptor Johann Gottfried Schadow in 1794. The gate and Quadriga, based on ancient Athenian temples, speaks to the aspirations of the Prussian rulers in their German empire. Perhaps it is not a stretch to say that its near-destruction at the end of World War II symbolized the end of the Nazis—a regime that went too far in its quest for power. The many

Figure 11.4
Brandenburg Gate, ca. 1958. "Soviet soldiers stand before Berlin's historic Branden-burg Gate, with its newly restored Quadriga statue on top. The statue was returned to the gate in June 1958, replacing the Soviet flag that flew there for years after the closing off of East Berlin. From the booklet 'A City Torn Apart: Building of the Berlin Wall.' For more information, visit the CIA's Historical Collections page." From https://www.flickr.com/photos/ciagov/8048135673

images of the Tor attest to its changing "stature" before and after the two world wars, and before and after the Berlin Wall (figure 11.4 and 11.5). It has been restored to its former magnificence, and it is a tourist magnet. As it is situated only steps away from several monuments, it is a metaphoric pres-ervation gateway. Its history is a powerful metaphor for the many meanings of *monumental*.

Figure 11.5
Brandenburger Tor today, 2017. Permission of Barbara G. Preece, photographer

"Memorial Way"

The Tor and the Reichstag are surrounded by memorials. Directly in front of the Reichstag is the installation "De Bevölkerung," ("To the People," or "To the Population") by Hans Haacke, in reference to the nearby inscription on the portico of the Reichstag, "Dem Deutschen Volke" ("[To] the German People"), which was placed there in 1916. By removing the word *Deutschen*, Haacke suggests that Berlin's population is broader than just German people. And perhaps he also seeks to remove the historical stench of the Third Reich, which resulted in widespread death and destruction. Directly next to it is the "Memorial to the Murdered Members of the Reichstag," which commemorates the Social Democratic and Communist delegates who were murdered by the National Socialists. It was designed by Dieter Appelt, Klaus W. Eisenlohr, Justus Müller, and Christian Zwirner. This piece also distinguishes itself from the Reichstag's neoclassical portico and majestic Corinthian columns by presenting us with dark, jagged cast-iron plates, the antithesis of the classical. The very structure of the Reichstag can be seen as having once belied order and reason. The jagged stones beckon us

Figure 11.6
Memorial to the Murdered Members of the Reichstag. Photograph by Michèle V. Cloonan, April 2016

to recall the dark history of the government between 1933 and 1945. Each plate contains, along the top edges, the names and birth and death dates of those who were murdered. The memorial can be extended if new names are discovered (figure 11.6).

Directly across the street are several large panels one of which has a quotation by the former Federal Chancellor, Helmut Schmidt: "The Nazi dictatorship inflicted a grave injustice on the Sinti and Roma. They were persecuted for reasons of race. These crimes constituted an act of genocide."

The panels offer a chronology of the genocide. Behind the fence on which the panels are attached, and at the edge of the Tiergarten, stands a reflecting pool: "Memorial to the Sinti and Roma of Europe Murdered under National Socialism." Around the edge of the pool is the poem "Auschwitz," by Roma poet Santino Spinelli. Triangular stones surround the pool, their shape suggesting the badges worn by concentration camp prisoners. In the middle of the pool is a triangular stone onto which a fresh flower is placed daily. The memorial was designed by Daniel (Dani) Karavan and opened in 2012.

Around the corner from the Sinti and Roma Memorial is a Memorial to Heinz Sokolowski (1917–1965), who was killed while attempting to escape over the Wall from East Berlin. This memorial looks rather makeshift: a tall, crude brown wooden cross with a photo of Sokolowski's body at the center. Behind the brown cross are white crosses with the names of others who tried unsuccessfully to escape the DDR. The memorial was put up in 1966, just a few feet away from the Berlin Wall. The fact that it is nearly contemporary to Sokolowski's murder makes it particularly effecting. It stands out in contrast to its neighboring memorials, which were planned for years and designed by prominent artists and architects (figure 11.7).

Close to the Sokolowski tribute is the Memorial to Homosexuals Persecuted under Nazism, a small structure designed by Michael Elmgreen and Ingar Dragset, the pair of artists whose work *Powerless Structure* appears in chapter 1 (see figure 1.8). The memorial was built in 2008. During World War II, as many as 15,000 gays[13] were deported to concentration camps where many of them died. At the street-side entry to the memorial is a long information board that informs us that some 50,000 homosexuals were convicted and sentenced to prison; "Lesbian women, too, were forced to conceal their sexuality. For decades, gays continued to be persecuted and prosecuted in both German post-war states and the homosexual victims of National Socialism were excluded from the culture of remembrance."[14] Efforts to establish a memorial to persecuted gay men and women began in 1992, at the same time as efforts were afoot to create memorials for the Sinti and Romas and the Jews. In 2002 the German Bundestag adopted a resolution to rehabilitate the victims of National Socialism and in 2003 it was decided to create a memorial. (See figure 11.8.)

The Holocaust-Denkmal, "Memorial to the Murdered Jews of Europe," which I described visiting earlier in the chapter, is the largest of the major memorials situated near the Reichstag. It consists of 2,711 large concrete stelae situated on an undulating concrete field that is spread over 4.7 acres. It was designed by Peter Eisenman and Richard Serra (who later pulled out) and completed in December 2004. The stelae represent the Jews who were murdered by the Nazis between 1933 and 1945. The memorial seems to reside under the watchful "eyes" of the American Embassy directly behind it, and the Reichstag two blocks further north, as can be seen in figure 11.9. The scale of the site is supposed to reflect the scale of the genocide; some six million Jews were murdered.

Figure 11.7
Memorial to Heinz Sokolowski. Photograph by Michèle V. Cloonan, April 2016

At the beginning of this chapter, I noted how each of these memorials was conceived of and created with extensive discussion and debate. While the notes from 2003 that were posted at the construction site of the Holocaust-Denkmal expressed a diversity of opinions, they were just one manifestation of the seventeen-year debate over the memorial. As Peter Schneider, chronicler of Berlin, described it:

[The debate] was passionate, wild, ambitious, banal, highly philosophical, grotesque and magnificent. There was no argument that wasn't made during the course of this debate, and none that wasn't just as quickly refuted. ... Jewish intellectuals explained that they didn't need the memorial and considered it completely superfluous or even harmful; others claimed that it was long overdue. Some non-Jewish spokespeople agreed with the former group, others with the latter.

Schneider further describes the three questions at issue: first, whether a memorial was needed, and second, would centralizing memory in Berlin detract attention from the concentration camps? The third question was whether the memorial should be dedicated to the Holocaust's Jewish

Figure 11.8
Memorial to Homosexuals (the image is projected on the inside), and exterior view, with rose. Photographs by Michèle V. Cloonan, April 2016

Figure 11.8 (continued)

Figure 11.9
Memorial to the Murdered Jews of Europe. Photographs by Michèle V. Cloonan, April 2016

Figure 11.9 (continued)

victims only, or include other persecuted groups. This last debate was resolved by the creation of other group-specific memorials.[15]

Once the decision was made to create a memorial site specifically for the Jewish victims, an equally diverse number of opinions was expressed about its design. How large should the memorial be? Should the monumentality of the crimes be reflected in the memorial? Proposed designs ranged from incorporating freight cars of the sort used to deport the Jews to creating a brick smokestack that would continually put out smoke. Still another proposal was to build a bus station from which buses would go to concentration camps and other sites of Jewish persecution.[16] In the end, Peter Eisenman and Richard Serra came up with the winning design.

Many views have been aired about the completed memorial. In Schneider's eyes, it is successful. He feels that the site blends in with the city's landscape and that it doesn't force a sense of guilt on visitors. Not everyone agrees. In an essay that is critical of the memorial, Richard Brody observed:

The title [of the memorial] doesn't say "Holocaust" or "Shoah"; in other words, it doesn't say anything about who did the murdering or why—there's nothing along the lines of "by Germany under Hitler's regime," and the vagueness is disturbing. Of course, the information is familiar, and few visitors would be unaware of it, but the assumption of this familiarity—the failure to mention it at the country's main memorial for the Jews killed in the Holocaust—separates the victims from their killers and leaches the moral element from the historical event, shunting it to the category of a natural catastrophe. The reduction of responsibility to an embarrassing, tacit fact that "everybody knows" is the first step on the road to forgetting.[17]

Brody implies that a memorial in and of itself is not necessarily an effective way to preserve history. Such sites must contain more context. While there is an underground information center, Brody does not believe that it is adequate. A look at some of the other sites in Berlin illustrates the many ways in which preservation can be viewed.

The Jewish Museum Berlin is another site of Jewish remembrance. One particular installation, Menashe Kadishman's *Shalechet (Fallen Leaves)* gives museum visitors a sensory experience of loss (figure 11.10). The following quote is from the Jewish Museum's website:

Visitors are encouraged to interact by walking on the exhibit itself: to see the open-mouths in terror, the faces of soundless screams; and to listen to the jarring clanging sounds when thick metal pieces jostle against other pieces.

It's an eerie atmosphere with the installation all to myself. I also feel what is unmistakably guilt as I tread on the "screaming" faces. Am I walking over representations

Figure 11.10
Detail of *Shalechet* (Falling Leaves), 1997–2001 by Menashe Kadishman (1932–2015).
Courtesy of the artist. Collection Jewish Museum Berlin

of living breathing [*sic*] people? I think these feelings are in fact necessary, that I need to have these feelings of loss. Something important has been taken away. It's as if the sculpture asks: "Germany is presently incomplete—will the country ever heal and be complete again?"[18]

Potsdamer Platz

Potsdamer Platz looks like a new place. (Though certainly not like just *any* place because sections of the Wall are situated throughout.) Not long after the Wall came down, major construction projects began such as the Sony Center, the Arkadin, and the skyscrapers of the Potsdamer Platz. The area had originally been a dynamic city center and the busiest traffic intersection in Europe in the 1920s[19]—but most of the buildings were destroyed during World War II. The neighborhood eventually became something of

Figure 11.11
Sections of the Berlin Wall. Photographs by Michèle V. Cloonan, April 2016

Figure 11.11 (continued)

a wasteland after the Berlin Wall went up because the East Germans tore down most of what had survived the war. Even buildings on the West Berlin side of the Platz were razed. There was a great demand to rebuild this part of Berlin once the Wall came down. With all of its shops, theaters, and restaurants, and because of its location near the old Checkpoint Charlie, this part of Berlin is once again lively (figure 11.11).

Shortly before the fall of the Wall in 1989, the CEO of the Daimler Group purchased fifteen thousand acres southwest of the Potsdamer Platz. His plan was to construct the Quartier Potsdamer Platz, with Renzo Piano as the architect. There were two impediments to building there: the marshy ground below, and the old Weinhaus Huth, a building that had been designated a landmark in 1979—solely because it had survived World War II and the subsequent demolitions in the area. It stood in the construction site area. The Daimler project manager decided to engage the press and the public into this enormous (and enormously complicated) project by inviting them as regular visitors to the operational headquarters, where a red information box was constructed as a viewing station. This was successful and may have led to later such efforts to include the public in building and restoration projects—like the Humboldt Box mentioned at the beginning of this chapter. The culmination of the PR activities for the Quartier was a concert at the 1996 groundbreaking with Daniel Barenboim conducting Beethoven's *Ode to Joy* while nineteen construction cranes moved their steel arms in time to the music.[20] The Weinhaus Huth has managed to fit in with its new neighbors.

On the other side of the Potsdamer Platz, the new Sony Center faced its own historic preservation challenge. The complex was designed by Helmut Jahn, a German-American architect. It is a spacious and light-flooded structure that has a large, suspended oval roof. However, Jahn needed to incorporate the ruins of the Kaisersaal, all that was left of the old Hotel Esplanade. According to a city guide: "The Senate of Berlin stipulated that Sony should preserve the 'Breakfast Room' and the 'Emperor's Hall' of the Grand Hotel Esplanade, both protected following the destruction in World War II [figure 11.12]. Accordingly, in 1996, the rooms were moved—1,300 tons were loaded onto wheels and shifted by 75 m (246 ft.) during the course of the week."[21]

The juxtaposition is strange; the Kaiseraal is encased in a glass box, perhaps to "match" the glass office tower next to it. Still, it looks rather as if it was plunked down on the site, a visitor from another planet. Is this historic preservation? In the literal sense it is; but it is now a building without context. On the other hand, the entire Potsdamer Platz is a modern re-creation of a neighborhood destroyed by war and the city divided by the Wall.

And what has happened to the Wall? Everything. Within a year after it began to come down on November 9, 1989, most of it had disappeared. Some of it was ruined during the takedown. Other parts were taken by tourists and scattered around the world. Still other sections were sold off. Some sections remain in the city, and there is a display of part of the Wall near the Wall Memorial on Bernauer Strasse. Pieces have also been refabricated and newly decorated.

As with all of the other memorials, there were many opinions about what should happen to the Berlin Wall. Some people who had to look at the Wall for many years wanted it taken down completely. Others felt that there needed to be a memorial, a reminder of Berlin as a divided city. Museums and memorials are scattered around the former city borders, which shows that Berliners have made a widespread, permanent record of the Wall by placing pieces of it in prominent places throughout the city.

Figure 11.12
Kaisersaal at the Sony Center. Photograph by Michèle V. Cloonan, April 2016

It is difficult to memorialize anything so soon after an event. That is a lesson about preservation that Berlin has had to grapple with over and over again. Still another: can a city have so many memorials and yet retain its vibrancy?

Yes it can. And one example is in the Kurfürstendam (see figure 11.13).

One of Berlin's most visible "mementos" of the destruction of World War II is the Kaiser-Wilhelm-Gedächtnis-Kirche (aka "the Ruined Church") in the Kurfürstendam, or Ku'damm, neighborhood. The Neo-Romanesque church was built in 1895, in honor of Wilhelm I, and was bombed in 1943. The ruins have been retained as a memorial. Originally, the church was to

Figure 11.13
Kurfürtendam, the "Ruined Church." Photograph by Michèle V. Cloonan, April 2016

Figure 11.14
Hiroshimastrasse. Photograph by Michèle V. Cloonan, April 2016

have been torn down for safety reasons. However, in a referendum, Berliners voted to preserve it, and the decision was made to build a new church next to it. Designed by Egon Eiermann, the new church was constructed from 1957 to 1963. The ruin is an arresting reminder of World War II, yet perhaps it is life-affirming as well. The Ku'damm is a lively neighborhood filled with shops, restaurants, and galleries. There is new construction under way around the church, a constant reminder that life goes on.

Another memorial worth mentioning is Hiroshimastrasse, because it reminds us of the international dimensions of war and peace. This little street is quite a contrast to the visible memorials all over Berlin (figure 11.14). It is in the Embassy district of Tiergarten and dates from 1862, when it was first named Hohenzollernstraße. In 1933 it was renamed Graf-Spee-Straße. In 1990 the street was renamed again after the first atomic bomb to fall on Japan. The Japanese Embassy was on the street during the Nazi era. When Berlin once again became the capital of Germany, the original embassy building was restored.[22]

Akira Kibi describes the street's current name:

The cool Hiroshima breeze can blow in any corners of the world connecting people with the wish of what Hiroshima stands for: the survival of human species and the respect for Nature. Such an idea already has some following: Berlin has a street, named Hiroshima Strasse, and the history of how this street came to be named after Hiroshima is very interesting, as told by Professor Mizushima of Waseda University.

Mr. Heinz Schmidt, a German school teacher who has visited Japan three times since 1965, made a bold proposal of changing the name of a park near his house in Berlin from its original name to Hiroshima Park. This attempt did not bear fruit the first time Mr. Schmidt tried it, in September of 1985, but his ardent effort continued for four years and it was paid off: he succeeded in converting the name of a bridge and a street that had until then been named after a Nazi Naval Admiral (those names represented Hitler's policy in 1933, to change names of every localities [sic] into militaristic names) into Hiroshima bridge and Hiroshima Strasse. His proposal passed the Berlin municipal assembly in April 1989, six months before the fall of the Berlin Wall. The street and the bridge were officially welcomed as legitimate member [sic] of the locales of Berlin in September, 1990; the moment Hiroshima's message gained its universality beyond national borders. The street just happens to run in Berlin connecting [the] Japanese embassy and that of Italy, resulting in the residential designation of current Japanese embassy as being at the number 6 [sic] of Hiroshima Street.[23]

The naming of this area after Hiroshima suggests, perhaps, that Berlin is becoming an international city of peace and reconciliation. In other words, the process of creating memorials has transformed the city from one in which the past is actively remembered to one in which that past can be transcended.

Demolition versus Reconstruction

One site in Berlin whose demolition was debated for years was the massive Palace of the Republic, the seat of the parliament of the German Democratic Republic in East Berlin. The complex was also a cultural center. The building was opened in 1976 and demolished from 2006 to 2008, so that the land could be used for a reconstruction of the Stadtschloss (also known as the Schloss [castle]). The demolition was opposed by former East Berliners as well as others who felt that this building was an important part of Berlin's history. Before demolition began, the building became a temporary center for art exhibitions and other events. The argument for tearing the building down was that it was filled with asbestos—though that ultimately caused the demolition to be extremely time consuming and costly. (One positive outcome was that 35,000 tons of steel were sent to Dubai where it was used to build the Burj Khalifa.)

What makes this demolition unusual is that the Stadtschloss—torn down by the East Germans in 1950—is being completely reconstructed from old plans, photographs, and other renderings. The Schloss was struck by Allied bombs in 1945 and lost its roof. Although it was damaged, the building was still structurally sound and it could have been restored. However, the East Germans probably viewed it as the ultimate symbol of Prussian imperial rule and thus wanted it destroyed. (Hitler saw it as "un-German" and never used it.)

The Schloss was first built in 1433, and it was modified over the centuries. Its history was tightly interwoven with Berlin's. Proponents of the reconstruction made the case that Berlin and the castle were one, while opponents pointed out some of its negative associations, like the fact that Kaiser Wilhelm II had declared World War I from its balcony, or that on November 9, 1919, Karl Liebknecht proclaimed the new German Socialist Republic. (Of course "negative" depends on one's own political beliefs. These two examples illustrate the many manifestations of the Schloss.) Others feared that the new castle would be an architectural pastiche. That it was also the center of the Revolution of 1848 may have swayed some. What may have convinced still others that the reconstruction would be beneficial is that the new palace would integrate the various buildings in the historical center of Berlin, which includes Museum Island, Berliner Dom, and the Lustgarten.

Schneider discusses Wilhelm von Boddien, who learned about the destruction of the Schloss when he was a student and who spent decades marshalling support for reconstructing it. In a story that is typical of preservation initiatives in Berlin, von Boddien found a unique way to garner support for reconstructing the castle: he commissioned French artist Catherine Feff to create a trompe l'oeil of the Berlin Schloss on more than one hundred thousand square feet of canvas. The canvas was mounted on scaffolding, to give "the illusion of a resurrection."[24] The effect was supposed to have been particularly striking at night, and the beautiful image seemed to have turned disbelievers of the reconstruction project into believers.

Construction of the new Schloss began in 2012; it is to be completed in 2019. The Berliner StadtSchloss will become the Berlin Palace Humboldt Forum (Humboldtforum), a global center for art and culture. This is one of the many "faces" of preservation—reconstruction—but not from the "leftovers" of a ruined building. It is actually not preservation at all in the strictest sense. With books and documents we can talk about facsimiles, which are generally copies made directly from originals. The Schloss is a copy, but made from original plans. Such an approach is a "re-creation" because the impetus is to bring something back. In this case it is the experience of the original castle.

Nefertiti and James Simon

At the beginning of this chapter I mentioned the Nefertiti bust that the Egyptians want back. She was discovered in 1912 at Tell al-Amarna by the German archeologist Ludwig Borchardt, in the workshop of the sculptor Thutmose (fl. 1350 B.C.E.). The Germans and Egyptians divided up the finds, but French officials were responsible for antiquities, even though Egypt was under British administration at the time (see chapter 4). Gustave Lefebvre was the Frenchman in charge. Borchardt may have concealed the value that he knew the Nerfertiti had. The bust went to Berlin where it was held in private hands until it was first displayed in the Neues Museum in 1924. It remained there until 1939, when Berlin museums were closed and their objects moved to secured spaces for safekeeping. Nefertiti was again moved and then found by the American army in 1945; she was then shipped to an American collecting point in Wiesbaden. The bust moved again, before being sent to West Berlin, where it was displayed at the Dahlem Museum. Then it was moved to the Egyptian Museum in Charlottenburg until 2009, when it was returned to the Neues Museum.[25] (It is nothing short of a miracle that the bust has survived so many moves and two world wars.)

Lefebvre's successor, Pierre Lacau, in 1925 set in motion the restitution claims that have continued. While he admitted to Lefebvre's mistake, he asserted that the Germans had essentially pulled a fast one. Nefertiti's "past" recently came to light again for two reasons: in 2011 the Egyptians called for its return, and in 2012 filmmaker Carola Wedel made a documentary about James Simon, the Jewish philanthropist who was one of the founders and financers of the Deutsche-Orient-Gesellschaft (German Orient Society). He had financed the excavation at Tell al-Amarna. Simon had been nearly forgotten in Germany despite his incredible largesse to German cultural institutions. He died in 1932, and because he was Jewish, his memory was expunged until recently.

Now, a new museum is being built on Museum Island in Simon's honor, and information about his collections is displayed in the Bode Museum. He connects back to Nefertiti in a fascinating way. He became the owner of the Nefertiti bust after the 1912 excavation, but he donated it to the Egyptian Museum of Berlin in 1920. However, he was willing to return the bust to Egypt in return for other artifacts from Egypt, believing that the restitution would ensure further excavations. The Egyptians have not given up hope that Nefertiti will return to Egypt. But it will not be likely to happen in the near future.

Rory McLean, an essayist who has written about Berlin for years, observed:

> That Germany is open and dynamic today is a consequence of taking responsibility for its history. In a courageous, humane and moving manner, the country is subjecting itself to a national psychoanalysis. This Freudian idea, that the repressed (or at least unspoken) will fester like a canker unless it is brought to the light, can be seen in Daniel Libeskind's tortured Jewish Museum, at the Holocaust Memorial and, above all, at the Topography of Terror. Be aware that this outdoor museum, built on the site of the former headquarters of the SS and Gestapo, is not for the fainthearted.[26]

Sometime in the near future, Berlin will address the preservation needs of its now large—and growing—multicultural population. Perhaps the city can draw on its many experiences to date. Public engagement will continue to be a foundation for preservation efforts.

* * *

I conclude this volume with this snapshot of Berlin. It is merely a snapshot, since it cannot tell the full story of this city's history, its rises and falls and subsequent rises, its attempts to rebuild itself after more than one (and more than one kind of) demolition, and its relationship to the theme of this book. Berlin's complexity as a city with a long and glorious (and also an inglorious) past—as a melting pot of peoples and ideas and embarrassments and victories and failures and destruction and reconstruction and transformation—raises a host of key questions. In no particular order: How long after a cataclysm should a city begin to rebuild itself? When a city has had many pasts, and when it is destroyed, which of its pasts do those who want to rebuild it choose? Should the city be rebuilt following past models or should it emerge with a fully new face? If part of the city's past reveals its inhumanity, its viciousness, and its embarrassments, whom should we follow: those who want to rebuild the city and hide those shortcomings or those who want to ensure they are remembered? What kinds of memorials should a city have?

How revealing should they be? Who should decide these things? And who should pay for the reconstruction and memorialization? We must remember that preservation is not merely of the built and natural world, it is of memory and emotion. And both kinds of preservation come with costs. There are many other questions like these that speak to preservation.

Berlin represents a host of issues I have examined throughout this book. I chose to end with the preservation activities of this great city because the kinds of issues that Berliners are dealing with today mirror many of the issues that we have looked at across these chapters. Tearing down buildings harkens back to the Richard Nickel chapter. The effects of war remind us of the chapters on Syria and cultural genocide. And even the difficult choices of what to save and what not to save are touched on in chapter 2, in my discussion of the sculptures of Soviet-era figures. In fact, nearly all of the themes covered in this book could in one way or another be related to Berlin.

What better place than Berlin to demonstrate all the ways in which preservation is monumental.

Notes

Preface

1. Jennifer Agiesta, "Poll: Majority Sees Confederate Flag as Southern Pride Symbol, Not Racist," *CNN*, July 2, 2015, http://www.cnn.com/2015/07/02/politics/confederate-flag-poll-racism-southern-pride/index.html (accessed October 11, 2017); Bre Payton, "Poll: Most Americans Don't Want Confederate Statues Torn Down," *The Federalist*, August 17, 2017, http://thefederalist.com/2017/08/17/poll-overwhelming-majority-americans-want-keep-confederate-statues (accessed September 17, 2017).

2. Christopher Carbone, "Oldest Historian Group on Confederate Monuments: Preservation and 'Historical Context' Needed," *Fox News*, August 31, 2017, http://www.foxnews.com/us/2017/08/31/oldest-historians-group-on-confederate-monuments-preservation-and-historical-context-needed.html (accessed September 17, 2017).

3. See https://www.conservation-us.org/docs/default-source/governance/position-paper-on-confederate-monuments-(september-2017).pdf?sfvrsn=2 (accessed September 17, 2017).

4. Carolyn E. Holmes, "Should Confederate Monuments Come Down? Here's What South Africa Did after Apartheid," *The Washington Post*, August 29, 2017, https://www.washingtonpost.com/news/monkey-cage/wp/2017/08/29/should-confederate-monuments-come-down-heres-what-south-africa-did-after-apartheid/?utm_term=.80e6105a9ce2 (accessed September 17, 2017).

5. Michèle V. Cloonan, "Monumental Preservation: A Call to Action," *American Libraries* 35, no. 8 (September 2004): 34–38.

6. David Lowenthal, *The Past Is a Foreign Country—Revisited* (Cambridge: Cambridge University Press, 2015).

Chapter 1

1. John Weever, *Ancient Funeral Monuments within the United Monarchie of Great Britain* (London, Thomas Harper, 1631), I B.

2. Ellen Pearlstein, "Conservation and Preservation of Museum Objects," in *Encyclopedia of Library and Information Sciences*, 3rd ed., ed. Marcia J. Bates and Mary Niles Maack (Boca Raton, FL: Taylor & Francis, 2010), 1269.

3. William D. Lipe, "A Conservation Model for American Archaeology," *The Kiva* 39, no. 3–4 (1974): 213.

4. Michèle Valerie Cloonan, ed., *Preserving Our Heritage: Perspectives from Antiquity to the Digital Age* (Chicago: Neal-Schuman 2015), xvii.

5. Salvador Muñoz Viñas, *Contemporary Theory of Conservation* (Amsterdam: Elsevier Butterworth-Heinemann, 2005), 10.

6. Ibid., 11.

7. For an excellent description of Carl Haber and IRENE see MacArthur Foundation, "Carl Haber: Audio Preservationist, *MacArthur Fellows Program,* September 25, 2013, https://www.macfound.org/fellows/class/2013 (accessed September 17, 2017).

8. *The Compact Edition of the Oxford English Dictionary*, vol. 1 (Glasgow: Oxford University Press, 1971), 1844.

9. Martial's *Epigrammaton*, Liber 10.2, quoted in Cloonan, *Preserving Our Heritage*, 7.

10. *Cassell's New Latin Dictionary* (New York: Funk & Wagnalls, 1960), 379.

11. For example, see André Schüller-Zwierlein, "Why Preserve? An Analysis of Preservation Discourses," *Preservation, Digital Technology & Culture* 44, no. 3 (2015): 98–122.

12. Alois Riegl, "The Modern Cult of Monuments: Its Essence and Its Development," trans. Karin Bruckner with Karen Williams, in *Historical and Philosophical Issues in the Conservation of Cultural Heritage*, ed. Nicholas Stanley Price, M. Kirby Talley, Jr., and Alessandra Melucco Vaccaro (Los Angeles: The Getty Conservation Institute, 1996), 69.

13. Sigmund Freud, "A Disturbance of Memory on the Acropolis: An Open Letter to Romain Rolland on the Occasion of His Seventieth Birthday." (The text was dated January 1936 by Freud.) See http://www.scribd.com/doc/6815599/A-Disturbance-of -Memory-on-the-Acropolis (accessed July 31, 2015). This text was no longer accessible in October 2017.

14. Ibid., 5.

15. Denise Scott Brown, Robert Venturi, and Steven Izenour, *Learning from Las Vegas* (Cambridge, MA: MIT Press, 1972). The book was revised in 1977, and a subtitle added: *The Forgotten Symbolism of Architectural Form.*

16. Ibid., 3.

17. Michèle V. Cloonan, "The Paradox of Preservation," *Library Trends* 56, no. 1 (Summer 2007): 133–147.

18. Judith Dupré, *Monuments: America's History in Art and Memory* (New York: Random House, 2007), 150.

19. The process is known as photo stencil grit-blasting.

20. "30th Anniversary of Vietnam Wall," USA Snapshots, *USA Today*, October 31, 2012, 1. In 2011, six million people visited the Lincoln Memorial and four million visited the Vietnam Veteran's Memorial, followed by the World War II Memorial (3.8 million), and the Korean War Veterans Memorial (3.1 million).

21. See http://www.vvmf.org/items (accessed September 9, 2016).

22. Dupré, *Monuments*, 221.

23. The statue is now at the Arken Museum of Art in Ishøj, Denmark.

24. Email communication from Roddy Williams, director of operations, NAMES Project/AIDS Memorial Quilt, March 26, 2016.

25. Dupré, *Monuments*, 160.

26. For more information see http://www.aidsquilt.org/about.

27. Richard Nickel and Aaron Siskind, with John Vinci and Ward Miller, *The Complete Architecture of Adler & Sullivan* (Chicago: Richard Nickel Committee, 2010).

28. Claudio Margottini, ed. *After the Destruction of Giant Buddha Statues in Bamiyan (Afghanistan) in 2001: A UNESCO's Emergency Activity for the Recovery and Rehabilitation of Cliff and Niches* (Berlin: Springer, 2014).

29. See Aryn Baker's "Afghanistan's Buddhas Can Be Rebuilt. But Should They? *Time*, March 2, 2011, http://world.time.com/2011/03/02/afghanistans-buddhas-can -be-rebuilt-but-should-they (accessed October 12, 2015).

30. Michael Falser. "The Bamiyan Buddhas, Performative Iconoclasm and the 'Image' of Heritage." In *The Image of Heritage: Changing Perception, Permanent Responsibilities*, ed. Simone Giometti, and Andrzej Tomaszewski, 157–169. Proceedings of the International Conference of the ICOMOS International Scientific Committee for the Theory and the Philosophy of Conservation and Restoration, 6–8 March 2009. Florence, Italy: ICOMOS, 2011.

31. Yamagata created a model of the laser Buddhas that he exhibited in Los Angeles in 2005. See "Afghanistan's Buddhas Rise Again," *The Atlantic*, June 10, 2015, http:// www.theatlantic.com/international/archive/2015/06/3d-buddhas-afghanistan /395576 (accessed November 11, 2016).

32. See for example https://www.google.com/search?q=Buddha+laser+projections &tbm=isch&tbo=u&source=univ&sa=X&ved=0ahUKEwj-iaeT2vnTAhXF5CYKHWhz Cs0Q7AkILg&biw=1879&bih=934 (accessed May 1, 2017).

33. Steven Davy, "They Were Destroyed by the Taliban. But Now the Giant Buddha Statues of Bamiyan Have Returned with 3-D Light Projection," *PRI's The World*, June 11, 2015, https://www.pri.org/stories/2015-06-11/they-were-destroyed-taliban-now -giant-buddha-statues-bamiyan-have-returned-3-d (accessed October 12, 2015).

34. András Riedlmayer, "Crimes of War, Crimes of Peace: Destruction of Libraries During and After the Balkan Wars of the 1990s," *Library Trends* 56, no. 1 (Summer 2007): 107–132.

35. See the ICOMOS website for more information about the heritage workers who have been murdered: http://www.icomos.org/en/178-english-categories/news /4084-icomos-supports-the-monuments-women-and-men-of-syria-and-iraq (accessed October 12, 2015).

36. Kareem Shaheen and Ian Black, "Beheaded Syrian Scholar Refused to Lead ISIS to Hidden Palmyra Antiquities," *The Guardian*, August 19, 2015, https://www.the guardian.com/world/2015/aug/18/isis-beheads-archaeologist-syria (accessed August 21, 2015).

37. Marcus Hall, *Earth Repair: A Transatlantic History of Environmental Restoration* (Charlottesville: University of Virginia Press, 2005), 245. See also, *Restoration and History: The Search for a Usable Environmental Past*, ed. Marcus Hall (London: Routledge, 2010).

38. John H. Stubbs, *Time Honored: A Global View of Architectural Conservation: Parameters, Theory, & Evolution of an Ethos* (Hoboken, NJ: John Wiley & Sons, 2009), 379–380.

39. In chapter 2, "Perspectives on Cultural Heritage," in my *Preserving Our Heritage: Perspectives from Antiquity to the Digital Age* (Chicago: Neal-Schuman, 2015), I present definitions of, and readings in, cultural heritage. See 19–57.

40. Frances A. Yates, *The Art of Memory* (Chicago: University of Chicago Press, 1966). Yates identifies the tradition of the "art of memory" as stemming from three Latin sources: The *Ad Herennium*, Cicero's *De oratore*, and Quintilian's *Institutio oratoria*.

41. Robert DeHart, "Cultural Memory," in *Encyclopedia of Library and Information Sciences*, 3rd ed., ed. Marcia J. Bates and Mary Niles Maack (Boca Raton, FL: CRC Press, 2010), 1363.

42. See Pierre Nora, *Realms of Memory (Les Lieux de Mémoire)*, https://en.wikipedia .org/wiki/Les_Lieux_de_Mémoire; accessed October 2016.

Chapter 2

1. "Book of Kells for Exhibition in US," *Irish Times* (February 5, 1976), 7. The "Troubles" referred to the ongoing political and social violence in Northern Ireland and elsewhere at the time. The early to mid-1970s was a particularly turbulent period, with

many acts of violence spilling over into the Republic of Ireland and mainland UK. The period is said to have ended with the Belfast Good Friday Agreement of 1998.

2. Other controversies included the selection of the items, the alleged political motivations of the exhibition, and the mediocre scholarship. For more information on these, see David H. Wright, "Shortchanged at the Met," *The New York Review of Books* 25, no. 7 (May 4, 1978): 32–34; and Wright, "Correspondence," *The New York Review of Books* 25, no. 12 (July 20, 1978): 47–48.

3. Liam de Paor, "The Book of Kells," *Irish Times*, February 24, 1976, 8.

4. The Metropolitan Museum of Art; the Fine Arts Museums of San Francisco, M.H. de Young Museum; Museum of Art, Carnegie Institute, Pittsburgh; Museum of Fine Arts, Boston; and the Philadelphia Museum of Art.

5. Memory of the World Register, Book of Kells, Ref. no. 2010–08, http://www .unesco.org/new/en/communication-and-information/memory-of-the-world/register /access-by-region-and-country/ie/ (accessed October 6, 2017).

6. Nathan Stolow, *Controlled Environment for Works of Art in Transit*, published for the International Centre for the Study of the Conservation of Cultural Property (London: Butterworths, 1966).

7. Nathan Stolow (1928–2014) remained active in the conservation field for most of the rest of his life.

8. G. Frank Mitchell, et al., *Treasures of Early Irish Art, 1500 BC–1500 AD* (New York: Metropolitan Museum of Art), 1977.

9. Caroline Walsh, "Art Treasures to Be Shown in US," *Irish Times*, June 30, 1977, 9.

10. Liam de Paor, "Irish Treasures for America," *Irish Times*, July 8, 1977, 8.

11. Muriel McCarthy, "Travelling Treasures," Letters to the Editor, *Irish Times*, July 13, 1977, 9.

12. Walsh, "Art Treasures to Be Shown in US."

13. Wright, "Shortchanged at the Met," 4.

14. Ibid., 2.

15. Ibid., 4.

16. "'Treasures of Early Irish Art': An Exchange," *The New York Review of Books* 25, no. 12 (July 20, 1978): 47–48.

17. Ireland_kells.doc-UNESCO, Memory of the World Register, Ref. no. 2010–08, section 4.1, http://www.unesco.org/new/en/communication-and-information/memory -of-the-world/register/access-by-region-and-country/ie/ (accessed October 6, 2017). Volume 2 of the Book of Kells, "The Gospel of Mark," was lent to the National

Gallery of Australia in Canberra, in 2000, and the volume suffered some damage. See *The Independent*, Friday, April 14, 2000, http://www.independent.co.uk/news/world /australasia/book-of-kells-damaged-on-australia-trip-279696.html (accessed June 21, 2015). Subsequently, as reported in the Memory of the World Register, the Board of Trinity College Dublin decided that the Book of Kells would no longer be lent.

18. For example, Vladimir Sorokin, "Let the Past Collapse on Time!" *The New York Review of Books*, May 8, 2014, http://www.nybooks.com/articles/archives/2014/may /08/let-the-past-collapse-on-time (accessed February 10, 2015).

19. Italics in the original. See http://en.wikipedia.org/wiki/Fallen_Monument_Park (accessed February 15, 2011).

20. Holograph diary of Michèle V. Cloonan, August 17–28, 1991, and holograph diary of Susan Swartzburg, August 16–27, 1991. Susan Swartzburg took notes in Moscow and created a word-processed document of which I have the photocopy that she sent me in fall 1991; she died in 1996.

21. From profile of Elena Baturina in *Forbes Magazine*, n.d., http://www.forbes.com /profile/elena-baturina (accessed December 1, 2016).

22. Deyan Sudjic, *Norman Foster: A Life in Architecture* (New York: Overlook Press, 2010), 240–246.

23. See http://www.atlasobscura.com/places/fallen-monument-park (accessed September 18, 2017).

24. Geoffrey A. Hosking and Yitzhak M. Brudny, *Reinventing Russia: Russian Nationalism and the Soviet State, 1953–1991* (Cambridge, MA: Harvard University Press, 2000), 69.

25. Geoffrey A. Hosking, "Memory in a Totalitarian Society: The Case of the Soviet Union," in *Memory: History, Culture and the Mind*, ed. Thomas Butler (Oxford: Basil Blackwell, 1989), 115.

26. Benjamin Forest and Juliet Johnson, "Unraveling the Threads of History: Soviet-Era Monuments and Post-Soviet National Identity in Moscow," *Annals of the Association of American Geographers* 92, no. 3 (September 2002): 524–547.

27. Ibid., 525.

28. Ibid., 536.

29. See http://en.wikipedia.org/wiki/Fallen_Monument_Park (accessed September 30, 2015).

30. Sherry Turkle, ed., *Evocative Objects: Things We Think With* (Cambridge, MA: MIT Press, 2007), 4–5. She cites as her influence Claude Levi-Strauss, *The Savage Mind*, and uses bricolage to mean material goods that conjure up ideas and memories.

31. Sorokin, "Let the Past Collapse on Time!"

32. Justinian A. Jampol, "Smashing Lenin Won't Save Ukraine," *New York Times*, March 4, 2014, Opinion and Editorial, http://www.nytimes.com/2014/03/04/opin ion/smashing-lenin-wont-save-ukraine.html (accessed June 21, 2015).

33. Ibid.

34. Sophia Kishkovsky, "Putin Dresses-down Culture Department over State of Monuments," *The Art Newspaper*, February 17, 2016, http://theartnewspaper.com /news/conservation/putin-s-dress-down-for-culture-department-over-state-of -monuments (accessed September 9, 2016; this site is no longer on the Web).

35. Zaryadye Park is an ambitious hybrid landscape design in Moscow that Diller Scofidio + Renfro are creating in collaboration with Hargreaves Associates, Citymak-ers, and an "international team of experts." Liza Premiyak, "A Wilderness in the City: How Diller Scofidio + Renfro's Zaryadye Park Could Help Fix Moscow," Arch-Daily, February 15, 2015, http://www.archdaily.com/598907/a-wilderness-in-the-city -how-diller-scofidio-renfro-s-zaryadye-park-could-help-fix-moscow (accessed September 18, 2017).

36. Kishkovsky, "Putin Dresses-down Culture Department over State of Monuments."

37. See http://russboroughhouse.ie/index.php/history/the-robberies (accessed June 22, 2015).

38. See Christie's press release: http://www.christies.com/about-us/press-archive /details?PressReleaseID=7887&lid=1&mob-is-app=false (accessed September 20, 2017).

39. See "The USSR's 1948 Instructions for the Identification, Registration, Mainte-nance, and Restoration of Architectural Monuments under State Protection," trans. Richard Anderson, in "Special Issue on the Preservation of Soviet Heritage," ed. Jean-Louis Cohen and Barry Bergdoll, *Future Anterior: Journal of Historic Preservation, History, Theory, and Criticism* 5, no. 1 (Summer 2008): 64–72, and in the same issue, "1973 Law of the Union of Soviet Socialist Republics: On the Protection and Use of Historic and Cultural Monuments," 74–80.

40. Cloonan, holograph diary, 26.

41. Swartzburg, holograph diary, photocopy of typescript. From the entry "Moscow, August 19–24," unpaginated.

Chapter 3

1. I have already mentioned the fire-bombing of Sarajevo's libraries.

2. John Donne, from *Devotions upon Emergent Occasions*, "Meditation XVII" (1624), http://triggs.djvu.org/djvu-editions.com/DONNE/DEVOTIONS/Download.pdf, 31 (accessed January 25, 2017).

3. András J. Riedlmayer, "Crimes of War, Crimes of Peace: Destruction of Libraries During and After the Balkan Wars of the 1990s," *Library Trends* 56, no. 1 (Summer 2007): 107.

4. Raphael Lemkin's (unfinished) history is contained in *Lemkin on Genocide*, ed. Steven Leonard Jacobs (Lanham, MD: Lexington Books, 2012), 59–401.

5. Fernando Báez, *A Universal History of the Destruction of Books: From Ancient Sumer to Modern Iraq*, trans. Alfred MacAdam (New York: Atlas & Co., 2004).

6. Rebecca Knuth, *Libricide: The Regime-Sponsored Destruction of Books and Libraries in the Twentieth Century* (Westport, CT: Praeger, 2003).

7. The term *human rights* probably came into use some time between Thomas Paine's *The Rights of Man* (1791) and William Lloyd Garrison's 1831 writings in *The Liberator*, in which he stated that he was trying to enlist his readers in "the great cause of human rights," although the term had been used by at least one author as early as 1742. See https://en.wikipedia.org/wiki/Human_rights (accessed June 30, 2015).

8. Sun Tzu, *The Art of War*, trans. Lionel Giles (Blacksburg, VA: Thrifty Books, 2009).

9. Keith E. Puls, ed., *Law of War Handbook*. JA 423. (Charlottesville, VA: The Judge Advocate General's Legal Center and School, 2005). See https://www.loc.gov/rr/frd /Military_Law/pdf/law-war-handbook-2004.pdf. (accessed September 20, 2017).

10. Dunant described the horrors of war in graphic detail in his book *Memory of Solferino*, which is in print more than 150 years after its publication. See Henry Dunant, *Memory of Solferino* (Geneva: International Committee of the Red Cross, 1862; repr. 1986).

11. Henri Dunant was awarded the Nobel Peace Prize in 1901. For additional biographical information about him see the Nobel Prize website: http://www.nobelprize .org/nobel_prizes/peace/laureates/1901/dunant-bio.html (accessed June 30, 2015).

12. For the full text of the *Lieber Code* see Avalon.law.yale.edu/19th_century/lieber .asp#sec2 (accessed June 28, 2015). This excerpt is from Article 35.

13. Michèle Valerie Cloonan, *Preserving Our Heritage: Perspectives from Antiquity to the Digital Age* (Chicago: Neal-Schuman, 2015), xxvi–xxvii.

14. The Convention for the Protection of Cultural Property in the Event of Armed Conflict is often referred to as the 1954 Hague Convention. In this chapter I use both forms of the name.

15. Puls, *Law of War Handbook*, 11.

16. U.S. Department of State, Office of the Historian. See https://history.state.gov /milestones/1921-1936/kellogg (accessed June 30, 2015).

17. Also referred to as UNCG, from the French version of the convention's name.

18. See http://www.icc-cpi.int/en_menus/icc/situationsandcases/pages/situationsand cases.aspx (accessed July 24, 2015).

19. The biographical information on Lemkin is based on the work of Jacobs (see note 4); John Cooper, *Raphael Lemkin and the Struggle for the Genocide Convention* (London: Palgrave Macmillan, 2008); and *The New York Times* obituary (August 30, 1959).

20. Thomas de Waal, "The G-Word: The Armenian Massacre and the Politics of Genocide," *Foreign Affairs* 94, no. 1 (January/February 2015): 142.

21. Cooper, *Raphael Lemkin and the Struggle for the Genocide Convention*, 18–19.

22. Quoted in William A. Schabas's introduction to Raphael Lemkin, *Axis Rule in Occupied Europe: Laws of Occupation, Analysis of Government, Proposals for Redress*, 2nd ed. (Clark, NJ: The Lawbook Exchange, 2008), ix.

23. Lemkin, *Axis Rule in Occupied Europe*, 79–95. The book was first published in 1944 in Washington, DC, by the Carnegie Endowment for International Peace, Division of International Law. This second edition is a reprint of the 1944 edition with another, new, introduction.

24. Jacobs, *Lemkin on Genocide*, 20.

25. Douglas Irvin-Erickson, "Genocide, The 'Family of Mind' and the Romantic Signature of Raphael Lemkin," *Journal of Genocide Research* 15, no. 3 (2013): 274.

26. Described on page 79 of Lemkin, *Axis Rule in Occupied Europe*.

27. See Cooper, *Raphael Lemkin and the Struggle for the Genocide Convention*. On p. 15, Cooper discusses how, when a student at the University of Lvov in Poland, Lemkin became interested in Volkgeist, the idea that every nation had a folk spirit, a unique culture. At the same time, he was researching Jewish pogroms and the murder of the Armenians. Thus, well before World War II and the rise of the Nazis, Lemkin was formulating ideas about genocide.

28. Ibid., 77.

29. For the complete text of the UNGC, see http://www.hrweb.org/legal/genocide .html (accessed July 2, 2015).

30. Dominik J. Schaller and Jürgen Zimmerer, eds., *Journal of Genocide Research* 7, no. 4 (2005): 441–578; and Donna-Lee Frieze, ed., *Journal of Genocide Research* 15, no. 3 (2013): 247–377.

31. The Final Report of the Commission of Experts Established Pursuant to Security Council Resolution 780. See United Nations, Security Council, "Letter Dated 24 May 1994 from the Secretary-General to the President of the Security Council," http://www.icty.org/x/file/About/OTP/un_commission_of_experts_report1994_en.pdf (accessed July 18, 2015).

32. See http://www.icty.org/x/cases/krstic/cis/en/cis_krstic_en.pdf for more information on this case.

33. Jacques Semelin, "Around the 'G' Word: From Raphael Lemkin's Definition to Current Memorial and Academic Controversies," *Genocide Studies and Prevention: An International Journal* 7, no. 1 (2012): 28.

34. Israel W. Charny cites many examples of genocide in his book *How Can We Commit the Unthinkable? Genocide: The Human Cancer* (Boulder, CO: Westview Press, 1982).

35. Katz, in Semelin, "Around the 'G' Word," 26.

36. Jaques Semelin, *Online Encyclopedia of Mass Violence*. See sciencespo.fr/ceri/en/ouvrage/oemv (accessed November 30, 2017).

37. *Merriam-Webster's 3rd New International Dictionary* (Springfield, MA: G. & C. Merriam, 1976), 2554.

38. Meghna Manaktala, "Defining Genocide," *Peace Review: A Journal of Social Justice* 24, no. 2 (2012): 185.

39. For example, Riedlmayer, "Crimes of War, Crimes of Peace"; Andrew Herscher and András Riedlmayer, *Destruction of Cultural Heritage in Kosovo, 1998–1999: A Postwar Survey* (Expert Report for the International Criminal Tribunal for the Former Yugoslavia). See also the Miloševic Trial Public Archive, *HRP Bard,* http://hague.bard.edu/reports/hr_riedlmayer-28feb2002.pdf (accessed November 30, 2017); and Andrew Herscher, *Violence Taking Place: The Architecture of the Kosovo Conflict* (Palo Alto, CA: Stanford University Press, 2010).

40. Shamiran Mako, "Cultural Genocide and Key International Instruments: Framing the Indigenous Experience," *International Journal on Minority and Group Rights* 19 (2012): 189. [The link cited in the article at footnote 55 was broken when consulted on July 18, 2015.]

41. Ibid., 175–194.

42. Ibid., 185.

43. United Nations Declaration on the Rights of Indigenous Peoples (2007), http://www.un.org/esa/socdev/unpfii/documents/DRIPS_en.pdf (accessed July 18, 2015).

44. See the UNESCO website, "Building Peace in the Minds of Men and Women: The Organization's History," n.d. http://www.unesco.org/new/en/unesco/about-us/who-we-are/history (accessed July 2, 2015).

45. Italics in the original. André Schüller-Zwierlein, "Why Preserve? An Analysis of Preservation Discourses," *Preservation, Digital Technology & Culture* 44, no. 3 (2015): 100.

46. Joseph Nye, *Soft Power: The Means to Success in World Politics* (New York: Public Affairs, 2004), p. 5.

47. See UNESCO, "Building Peace in the Minds of Men and Women: Introducing UNESCO: What We Are," n.d., http://www.unesco.org/new/en/unesco/about-us /who-we-are/introducing-unesco (accessed July 22, 2015).

48. See "News: United Kingdom," *The Art Newspaper*, no. 269 (June 2015): 10. It was ratified on September 12, 2017.

49. "Together with a number of other countries, the UK Government did not ratify the [c]onvention when it was first drafted because it considered that it did not provide an effective regime for the protection of cultural property. However, the introduction of the Second Protocol removed this concern and in 2004 a commitment to ratify the convention was announced." See DCMS, "Hague Convention 1954," Department for Culture, Media and Sport, http://old.culture.gov.uk/what_we_do /cultural_property/6630.aspx (accessed July 18, 2015).

50. UNESCO, "Armed Conflict and Heritage: 1999 Second Protocol to the Hague Convention," n.d., http://www.unesco.org/new/en/culture/themes/armed-conflict -and-heritage/convention-and-protocols/1999-second-protocol (accessed September 18, 2017).

51. Ibid.

52. See the Heritage for Peace website at http://www.heritageforpeace.org/syria -culture-and-heritage (accessed July 11, 2015). There will be more about international on-the-ground preservation projects in the following chapters.

53. Irina Bokova, "Discours de la Directrice générale de l'UNESCO Irina Bokova, à l'occasion de la Réunion de haut niveau pour la protection du patrimoine culturel syrien," August 29, 2013 (DG/2013/119). [French and English]. See http://www .unesco.org/new/fileadmin/MULTIMEDIA/HQ/ERI/pdf/ReunionPatrimoineSyrie.pdf (accessed July 3, 2015).

54. Raphael Lemkin, "Genocide—A Modern Crime," *Free World* (April 1945): 39.

Chapter 4

1. Permission granted by the author, who holds the copyright. Published in *Sukoon* 3, no. 2 (Summer 2015) at http://www.sukoonmag.com/responsive/wp-content /uploads/Sukoon-Mag-Issue-6-S-2015.pdf (accessed October 11, 2017).

2. An excellent source is Syria Untold, "an independent digital media project exploring the storytelling of the Syrian struggle and the diverse forms of resistance. We are a team of Syrian writers, journalists, programmers and designers living in the country and abroad trying to highlight the narrative of the Syrian revolution, which Syrian men and women are writing day by day." See http://www.syriauntold.com /en/about-syria-untold (accessed August 15, 2016).

3. The sources I cite use the terms equally. Less commonly used is *Daesh* or *Da'ish*, the acronym in Arabic. The group IS/ISIL/ISIS uses *al-Sham* instead of Syria. Al-Sham refers to ancient Syria, a large region in which people could travel freely, not post–World War I nation-state Syria with its smaller borders. It is likely that ISIS uses al-Sham instead of Syria because it disputes Syria's current borders.

4. This was widely reported in the news. My sources included *The Guardian*, https://www.theguardian.com/world/2015/aug/18/isis-beheads-archaeologist-syria (accessed August 2016); and *The Art Newspaper,* http://theartnewspaper.com/news/news/italy-s-museums-honour-archaeologist-murdered-by-isil (accessed August 2016).

5. Robin Yassin-Kassab and Leila Al-Shami, *Burning Country: Syrians in Revolution and War* (London: Pluto Press, 2016), 1. According to UNESCO World Heritage Site information, "Aleppo has stood at the crossroads of trade routes across Syria since at least the 3rd millennium B.C., when the city was first mentioned in ancient Syrian manuscripts"; see UNESCO, "Aleppo," *Silk Road: Dialogue, Diversity & Development*, http://en.unesco.org/silkroad/content/aleppo (accessed August 15, 2016).

6. See *Encyclopaedia Britannica*, "The Phoenician Alphabet," https://www.britannica.com/topic/Phoenician-alphabet (accessed August 2, 2016).

7. The souk was largely destroyed in September and October of 2012; for an account of it, see Kevin Rushby, "Remembering Syria's Historic Silk Road in Aleppo," *The Guardian*, October 5, 2012, https://www.theguardian.com/travel/2012/oct/05/aleppo-souk-syria-destroyed-war (accessed August 17, 2016).

8. Yassin-Kassab and Al-Shami, *Burning Country*, 1–2.

9. For a thorough and engaging account of Sir Mark Sykes and François Georges Picot—and for the unfortunate legacy of their work that would follow—see M. E. McMillan, *From the First World War to the Arab Spring: What's Really Going on in the Middle East?* (New York: Palgrave Macmillan, 2016), 69–75.

10. Ayse Tekdal Fildis, "The Troubles in Syria: Spawned by French Divide and Rule," *Middle East Policy Council* 18, no. 4 (Winter 2011): 1, http://www.mepc.org/troubles-syria-spawned-french-divide-and-rule (accessed August 18, 2016).

11. Ibid., 2.

12. Ibid., 1; and Adeed I. Dawisha, *Syria and the Lebanese Crisis* (New York: St. Martin's Press, 1980), 17.

13. Kerim Yildiz, *The Kurds in Syria: The Forgotten People* (London: Pluto Press, 2005), 14.

14. Ibid., 23.

15. Fildis, "The Troubles in Syria," 3–6.

16. Ibid. 6.

17. McMillan, *From the First World War to the Arab Spring*, 83.

18. Craig S. Davis, *The Middle East for Dummies* (Hoboken, NJ: Wiley Publishing, 2003), 257.

19. Yassin-Kassab and Al-Shami, *Burning Country*, 13.

20. Ibid. 198.

21. *Arab Spring* refers to the wave of protests—violent and peaceful—that started in Tunisia in December 2010. It soon spread to other countries in the Arab League. In Egypt, the protests were put down. The initial wave of protests had subsided in most places by mid-2012. Thus far, only Tunisia has transitioned to a democratic government.

22. Yassin-Kassab and Al-Shami, *Burning Country*, 36.

23. Ibid., 51.

24. See, for example, Jonathan Steele, "The Syrian Kurds Are Winning," *The New York Review of Books* 62, no. 19 (December 3, 2015) (review of Michael M. Gunter, *Out of Nowhere: The Kurds of Syria in Peace and War* [London: Hurst, 2015]), http://www.nybooks.com/articles/2015/12/03/syrian-kurds-are-winning (accessed August 19, 2016).

25. Nikolaos van Dam, *The Struggle for Power in Syria: Politics and Society under Asad and the Ba'th Party* (London: I. B. Tauris, 1997), 1. (I used the 1996 edition, repr. in 1997, to draw comparisons between the mid-1990s and now.)

26. Ibid., 1.

27. See BBC News, "Life and Death in Syria: Five Years into War, What Is Left of the Country?" March 15, 2016, http://www.bbc.co.uk/news/resources/idt-841ebc3a -1be9-493b-8800-2c04890e8fc9 (accessed August 18, 2016).

28. Max Fisher, "The One Map that Shows Why Syria Is So Complicated," *The Washington Post*, August 27, 2013, https://www.washingtonpost.com/news/world views/wp/2013/08/27/the-one-map-that-shows-why-syria-is-so-complicated (accessed August 19, 2016).

29. Ibid.

30. Giorgio Cafiero, "Hamas and Hezbollah Agree to Disagree on Syria," *Huffington Post* (January 31, 2014; updated April 2, 2014), http://www.huffingtonpost.com/gior gio-cafiero/hamas-and-hezbollah-agree_b_4698024.html (accessed August 15, 2016).

31. BBC News, "Life and Death in Syria: Five Years into War, What Is Left of the Country?" March 15, 2016, http://www.bbc.co.uk/news/resources/idt-841ebc3a -1be9-493b-8800-2c04890e8fc9 (accessed August 24, 2016).

32. *Lemkin on Genocide*, ed. by Steven Leonard Jacobs (Lanham, MD: Lexington Books, 2012), 59–401. Page x explains Lemkin's originally proposed text.

33. Raphael Lemkin, *Axis Rule in Occupied Europe: Laws of Occupation, Analysis of Government, Proposals for Redress*. 2nd ed. (Clark, NJ: The Lawbook Exchange, 2008). The book was first published in 1944 by the Carnegie Endowment for International Peace, Division of International Law.

34. *Lemkin on Genocide*, 3 and 4.

35. Lemkin, *Axis Rule in Occupied Europe*, xviii.

36. Lemkin sought support for his research, and his book proposal is included in *Lemkin on Genocide*, ed. Steven Leonard Jacobs.

37. Ibid., 3.

38. Karen Leigh, "Hope for Palmyra's Future," *The Wall Street Journal*, April 5, 2016, Arts in Review, D5.

39. From the International Committee of the Blue Shield website: http://www.ancbs .org/cms/en/about-us/about-icbs (accessed August 28, 2016).

40. Peter Stone, "War and Heritage: Using Inventories to Protect Cultural Property," *The Getty Conservation Newsletter* 28, no. 2 (Fall 2013), http://www.getty.edu/conser vation/publications_resources/newsletters/28_2/war_heritage.html (accessed August 29, 2016).

41. Marie-Thérèse Varlamoff and George MacKenzie, "Archives in Times of War: The Role of IFLA and ICA within ICBS (International Committee of the Blue Shield)," in *A Reader in Preservation and Conservation, IFLA Publications 91*, ed. Ralph W. Manning and Virginie Kremp (München: K. G. Saur, 2000; repr. 2013), 151.

42. Ibid., 154–155.

43. ICBS, The Seoul Declaration on the Protection of Cultural Heritage in Emergency Situations, http://icom.museum/uploads/media/111210_ICBS_seoul_declara tion_final.pdf (accessed August 28, 2016).

44. For more information see https://en.unesco.org/syrian-observatory (accessed August 28, 2016).

45. Michèle V. Cloonan, Allison Cuneo, and Father Columba Stewart, "Update on Syria," *Preservation, Digital Technology & Culture* 45, no. 2 (2016), DOI 10.1515 /pdtc-2016-0015. See also the HMML website, http://www.hmml.org/about-us.html (accessed August 28, 2016).

46. Makenna Murray, "Building Digital Preservation Capacity in the Middle East through Training: Discussing Training in Lebanon and the Future of Project Anqa with CyArk Field Manager, Ross Davison," July 28, 2016, http://www.cyark.org

/news/building-digital-preservation-capacity-in-the-middle-east-through-training (accessed August 28, 2016).

47. Sarah Parcak, "Archaeological Looting in Egypt: A Geospatial View (Case Studies from Saqqara, Lisht, and el Hibeh)," *Near Eastern Archaeology* 78, no. 3 (2015): 196–203.

48. Kurt Prescott, "Seven Things You Should Know about ASOR's Syrian Heritage Initiative," n.d. American School of Oriental Research. http://asorblog.org/2014/09/10/6-things-you-need-to-know-about-asors-syrian-heritage-initiative (accessed August 24, 2016).

49. Jeanie Chung, "In the Fray," interview with Matthew Barber, *The University of Chicago Magazine* 108, no. 2 (Winter 2016): 21.

Chapter 5

1. G. Thomas Tanselle, "A Rationale of Collecting," *Studies in Bibliography* 51 (1998): 1.

2. From the Getty Research Institute (GRI), n.d., "Collection Inventories and Finding Aids," http://archives2.getty.edu:8082/xtf/view?docId=ead/890164/890164.xml ;query=;brand=default (accessed September 26, 2016).

3. One could write an entire essay about the *types* of works there are on collecting. Perhaps the most common are the memoirs of collectors, including the many examples written by A. Edward Newton. Holbrook Jackson looked at the "causes" of book collecting in *The Anatomy of Bibliomania* (London: Soncino Press, 1930). Nicholas A. Basbanes wrote a book of portraits about collectors, *A Gentle Madness: Bibliophiles, Bibliomanes, and the Eternal Passion for Books* (New York: Henry Holt & Co., 1995). Thomas Frognall Dibdin studied "book-madness" in his famous book *The Bibliomania*, of which there are several editions, the first published in 1809. Art historian and psychoanalyst Werner Muensterberger wrote about the conscious and subconscious motivations for collecting in his *Collecting: An Unruly Passion: Some Psychological Perspectives* (Princeton: Princeton University Press, 1994). Many well-known scholars have also weighed in on collecting such as William James and Walter Benjamin. John Carter examined collectors in *Taste and Technique in Book Collecting: A Study of Recent Developments in Great Britain and the United States* (New York: R. R. Bowker, 1948) and William Rees Mogg wrote a work on purchasing rare books, *How to Buy Rare Books: A Practical Guide to the Antiquarian Book Market* (Oxford: Phaidon, Christies, 1985). See also Sidney E. Berger, *Rare Books and Special Collections* (Chicago: Neal-Schuman, 2014) for more on collectors and collecting.

4. Tanselle, "A Rationale of Collecting," 1.

5. In an email from September 6, 2016, Marcia Reed reported that the books are still there, primarily for the use of the GRI Provenance Index staff.

6. OCLC is the world's largest bibliographic service, recording the holdings of about 17,000 libraries. It calls itself a "global library cooperative" and has records for more than three billion items. See https://www.oclc.org/en/home.html (accessed September 7, 2017).

7. Sherry Turkle, ed. *Evocative Objects: Things We Think With* (Cambridge, MA: MIT Press, 2007).

8. Ibid., 307.

9. See the long definition of *ephemera* at the Ephemera Society of America website: http://www.ephemerasociety.org/def.html (accessed September 28, 2016).

10. See the description of the Shelburne Museum at https://en.wikipedia.org/wiki /Shelburne_Museum (accessed September 20, 2016).

11. Florence Fearrington, *Rooms of Wonder: From Wunderkammer to Museum, 1599–1899: An Exhibition at the Grolier Club, 5 December 2012–2 February 2013.* New York: The Grolier Club, 2012.

12. Ibid., 12.

13. For more details, see Museum of Fine Arts, "Kunstkammer Gallery," http://www .mfa.org/collections/featured-galleries/kunstkammer-gallery?utm_source=google -grant&utm_medium=cpc&utm_content=gallery-kunstkammer_4&utm_cam paign=mfa-brand&gclid=CNz72IXwss8CFQ9ahgodqCUCFw (accessed September 27, 2016).

14. This statement is on the dust jacket of *Portrait of an Obsession*, Barker's adaptation of Munby's *Phillipps Studies* (see note 15).

15. A. N. L. Munby wrote a multi-volume work about Phillipps: *Phillipps Studies,* 5 vols. (Cambridge: Cambridge University Press, 1951–1960). Nicolas Barker adapted Munby's book into the highly readable *Portrait of an Obsession* (New York: G. P. Putnam's Sons, 1967). For a brief account of Phillipps, see also Basbanes, *A Gentle Madness* (see note 3).

16. Basbanes, *A Gentle Madness.*

17. See the Library of Congress website: https://www.loc.gov/exhibits/jefferson/jef flib.html (accessed September 26, 2016).

18. Thomas Jefferson, [Letter] "From Thomas Jefferson to Ebenezer Hazard, 18 February 1791," https://founders.archives.gov/documents/Jefferson/01-19-02-0059 (accessed September 18, 2017).

19. See Kurtis R. Schaeffer, *The University of Virginia's First Library: A Guide to Researching the First Library of the University of Virginia Using the 1828 Catalogue.* See http://www.viseyes.org/library/ResearchGuide.pdf (accessed September 26, 2016).

20. Carter, *Taste and Technique*, 3.

21. Ibid., 4.

22. Again, there is a large literature. Lawrence W. Towner, *An Uncommon Collection of Uncommon Collections: The Newberry Library* (Chicago: The Newberry Library, 1970) and Kenneth Clark, *One Hundred Details from Pictures in the National Gallery* (London: The National Gallery, 1938) are two "old-school" descriptions of institutional collecting. Today many scholars and professional leaders have reassessed how, why, and what we collect. See for example James Cuno, *Who Owns Antiquity? Museums and the Battle over Our Ancient Heritage* (Princeton: Princeton University Press, 2008) and Susan M. Pearce et al., *The Collector's Voice: Critical Readings in the Practice of Collecting* (Aldershot, England: Ashgate, 2002).

23. *New York Times*, "Jean Brown 82, Avid Collector of Dada, Surrealism and Fluxus," obituary, May 4, 1994, http://www.nytimes.com/1994/05/04/obituaries/jean-brown -82-avid-collector-of-dada-surrealism-and-fluxus.html (accessed November 20, 2016).

24. Many thanks to Jonathan Brown, Jean Brown's son and a distinguished art historian. His memoir, *In the Shadow of Velázquez: A Life in Art History* (New Haven: Yale University Press, 2014), contains a charming description of his parents as collectors.

25. From the Jean Brown Collection finding aid at http://archives2.getty.edu:8082 /xtf/view?docId=ead/890164/890164.xml;chunk.id=ref15;brand=default (accessed September 28, 2016).

26. Brown, *In the Shadow of Velázquez*, 7–8.

27. For a full discussion of these issues, see Richard Rinehart and Jon Ippolito, *Re-Collection: Art, New Media, and Social Memory* (Cambridge, MA: MIT Press, 2014), 82–83, 104.

28. Tanselle, "A Rationale of Collecting," 4.

29. See Albert B. Lord, *Singer of Tales* (Cambridge, MA: Harvard University Press, 1960), esp. chapter 6. Even a "performance" by one tale teller may differ from that same person's next performance, depending on many things: setting, audience, time of day or year, the amount of sleep the tale teller has had, the amount of time she or he is given to tell the tale, and so on.

30. See UNESCO, "What Is Intangible Cultural Heritage," https://ich.unesco.org/en /what-is-intangible-heritage-00003 (accessed October 5, 2017).

31. Ibid.

Chapter 6

1. Richard Cahan, *They All Fall Down: Richard Nickel's Struggle to Save America's Architecture* (New York: John Wiley & Sons, 1994), 149. See also Richard Cahan and

Michael Williams, *Richard Nickel's Chicago: Photographs of a Lost City* (Chicago: City-Files Press, 2008).

2. John H. Stubbs, *Time Honored: A Global View of Architectural Conservation: Parameters, Theory, & Evolution of an Ethos* (Hoboken, NJ: John Wiley & Sons, 2009), 159.

3. Ibid., 162.

4. Cahan, *They All Fall Down*, 101; Stubbs, *Time Honored*, 171.

5. The *Chicago School* is a loose term that has been used to refer to a number of architects associated with Chicago from around 1880 through the first quarter of the twentieth century. Daniel Bluestone traces its early use to Harriet Monroe (in 1912) and H. L. Mencken (in 1925). See Bluestone's *Buildings, Landscapes, and Memory: Case Studies in Historic Preservation* (New York: W. W. Norton, 2011), 165.

6. The Richard Nickel Archive contains a large number of photographs of buildings by William LeBaron Jenney, Daniel Burnham and John Wellborn Root (Burnham & Root), and John Holabird and Martin Roche (Holabird & Roche).

7. Art Institute of Chicago Press Release, "Archive of Architectural Photographer Richard Nickel Goes to Art Institute's Ryerson and Burnham Libraries," November 17, 2010, http://www.artic.edu/sites/default/files/press_pdf/Richard_Nickel_Archi .pdf (accessed February 28, 2016).

8. John Vinci played an important role. See "John Vinci: The Modern Preservationist," *Architects + Artisans: Thoughtful Design for a Sustainable World*, October 8, 2014, http://architectsandartisans.com/index.php/2014/10/john-vinci-the-modern -preservationist (accessed February 9, 2016).

9. Cahan, *They All Fall Down*, 26.

10. See the home page of the Chicago Architecture Foundation, *Buildings of Chicago*, http://www.architecture.org/architecture-chicago/buildings-of-chicago (accessed February 28, 2016).

11. Max Page and Randall Mason, eds. "Introduction: Rethinking the Roots of the Historic Preservation Movement," in *Giving Preservation a History: Histories of Historic Preservation in the United States* (New York and London: Routledge, 2004), 7.

12. Robert E. Stipe, ed. *A Richer Heritage: Historic Preservation in the Twenty-first Century* (Chapel Hill: University of North Carolina Press, 2003), viii and ix.

13. Page and Mason, "Introduction: Rethinking the Roots of the Historic Preservation Movement," 14.

14. Ibid., 11; and Lee Bey, "The Storied Mecca Apartments Live Again, Thanks to 'Life' Magazine Archives," June 21, 2012, http://www.wbez.org/blogs/lee-bey/2012 -06/storied-mecca-apartments-live-again-thanks-life-magazine-archives-100301 (accessed February 18, 2016).

15. See the National Park Service site: http://www.nps.gov/hdp/habs (accessed February 8, 2016).

16. Bluestone, *Buildings, Landscapes, and Memory,* 170.

17. Daniel Bluestone, "Chicago's Mecca Flat Blues," in *Giving Preservation a History: Histories of Historic Preservation in the United States,* ed. Max Page and Randall Mason (New York and London: Routledge, 2004), 208.

18. Arnold R. Hirsch, "Urban Renewal," *Encyclopedia of Chicago,* http://www.ency clopedia.chicagohistory.org/pages/1295.html (accessed March 5, 2016).

19. Ibid.

20. Ibid.

21. This brief biography is drawn from WTTW, "The Richard Nickel Story," *Chicago Stories,* 2002, http://interactive.wttw.com/a/chicago-stories-richard-nickel-story (accessed January 28, 2017); and notes from the Richard Nickel Archive at the Art Institute of Chicago. Additional information was taken from the "Richard Nickel" posting in *Wikipedia,* https://en.wikipedia.org/wiki/Richard_Nickel (accessed February 8, 2016).

22. WTTW, "The Richard Nickel Story."

23. Barbara Sciacchitano, "Historic Preservation," *Encyclopedia of Chicago,* http://www.encyclopedia.chicagohistory.org/pages/586.html (accessed February 6, 2016).

24. Ibid.

25. ICOMOS. International Charter for the Conservation and Restoration of Monuments and Sites (The Venice Charter 1964). https://www.icomos.org/charters /venice_e.pdf (accessed January 28, 2017); Benjamin Schwarz, "The Architect of the City," *The Atlantic,* March 2011, http://www.theatlantic.com/magazine/archive /2011/03/the-architect-of-the-city/308389 (accessed February 5, 2016).

26. Richard Nickel and Aaron Siskind, with John Vinci and Ward Miller, *The Complete Architecture of Adler & Sullivan* (Chicago: Richard Nickel Committee, 2010).

27. Sage Stossell, "The Architecture of Louis Sullivan: A Photo Gallery," *The Atlantic,* February 8, 2011, http://www.theatlantic.com/entertainment/archive/2011/02/the -architecture-of-louis-sullivan-a-photo-gallery/70108/-slide1 (accessed February 5, 2016).

28. Quoted in Cahan, *They All Fall Down,* 89.

29. WTTW, "The Richard Nickel Story." The excerpt from Daley's speech, available at http://video.wttw.com/video/2365335740/, occurs from 9:34–9:46.

30. Cahan, *They All Fall Down,* 130.

31. Ibid., 26.

32. Hugh Morrison, *Louis Sullivan: Prophet of Modern Architecture* (New York: Museum of Modern Art and W. W. Norton, 1935).

33. Bluestone, *Buildings, Landscapes, and Memory*, 177. His source for Le Corbusier: "Views," letters by Thomas B. Stauffer and Le Corbusier on the Garrick Theater in Chicago, published in *Progressive Architecture* 42, no. 6 (June 1961): 208.

34. Sarah Rogers Morris, "Richard Nickel's Photography: Preserving Ornament in Architecture," *Future Anterior: Journal of Historic Preservation, History, Theory, and Criticism* 10, no. 2 (Winter 2013): 79.

35. Holly Giermann, "The Destruction of a Classic: Time-Lapse Captures Demolition of Chicago's Prentice Women's Hospital," February 4, 2015, http://www.arch daily.com/595078/the-destruction-of-a-classic-time-lapse-captures-demolition-of -chicago-s-prentice-women-s-ospital (accessed February 4, 2016).

36. Page and Mason, "Introduction: Rethinking the Roots of the Historic Preservation Movement," 5.

37. Studs Terkel, *Stud Terkel's Chicago* (New York: The New Press, 1986), 75–76.

Chapter 7

1. Hillel Schwartz, *The Culture of the Copy: Striking Likenesses, Unreasonable Facsimiles* (New York: Zone Books, 1996), 211.

2. Ibid., 214.

3. Christoph Scheiner, *Phantographice seu Ars delineandi res quaslibet per parallole-grammum lineare* (Roma: Grignani, 1631). For a modern description, see Sidney E. Berger, *The Dictionary of the Book: A Glossary for Book Collectors, Booksellers, Librarians, and Others* (Lanham, MD: Rowman & Littlefield, 2016), 182.

4. Barbara J. Rhodes and William W. Streeter, *Before Photocopying: The Art and History of Mechanical Photocopying, 1780–1938*. New Castle, DE: Oak Knoll, 1999.

5. Berger, *The Dictionary of the Book*, see "Typescript," 266; "Carbon Paper" posting in *Wikipedia*, https://en.wikipedia.org/wiki/Carbon_paper (accessed October 29, 2016).

6. Southern Regional Library Facility, University of California, "Chronology of Microfilm Developments, 1800–1900." *The History of Microfilm: 1839 to the Present*, https://www.google.com/webhp?tab=mw&ei=RDcWWNPnPInumQGEqqbYCA &ved=0EKkuCAUoAQ-q=microfilm+timeline (accessed October 29, 2016).

7. The word itself is derived from the Greek for *dry* and *writing*. See Berger, *The Dictionary of the Book*, 288.

8. Cornell University Law School, *Federal Rules of Evidence*, Article X, "Contents of Writings, Recordings, and Photographs," https://www.law.cornell.edu/rules/fre /article_X (accessed October 31, 2016). See Rule 1003, Admissibility of Duplicates.

9. Ibid. See Rule 1005, "Copies of Public Records to Prove Content."

10. This is different from electronic authentication (i.e., e-authentication) which is used to establish user identities by means of tokens, passwords, "challenge questions," biometrics, and so on.

11. See https://archive.org/scanning (accessed October 30, 2016).

12. Paul Conway, "Preserving Imperfection: Assessing the Incidence of Digital Imaging Error in HathiTrust," *Preservation, Digital Technology & Culture* 42, no. 1 (2013): 17, 19.

13. Paul Duguid, "Inheritance and Loss: A Brief Survey of Google Books," *First Monday* 12, no. 8 (August 6, 2007). See http://firstmonday.org/ojs/index.php/fm /article/view/1972/1847 (accessed October 27, 2016).

14. Ibid.

15. "Because the Contractor is responsible for accurately and thoroughly inspecting their work and identifying and correcting all errors for which they are responsible, NLM expects to receive shipments containing no filming errors," in National Library of Medicine, *NLM Microfilming Project, Statement of Work*, revised May 17, 2001, 43. See https://www.nlm.nih.gov/psd/pres/contracts/filmsowfy01.txt (accessed October 30, 2016). See also Karen Coyle, "Mass Digitization of Books," *Journal of Academic Librarianship* 32, no. 6 (November 2006): 643.

16. Conway, Preserving Imperfection," 27.

17. A related issue is whether cameras have become a replacement for experiencing art, natural landscapes, and so on. See Larry Rosen, "Is It Live or Is It Memorex? Viewing the World through a Camera Lens," *Psychology Today,* July 18, 2013, https:// www.psychologytoday.com/blog/rewired-the-psychology-technology/201307/is-it -live-or-is-it-memorex (accessed October 20, 2016).

18. Schwartz, *The Culture of the Copy*, 212.

Chapter 8

1. Italics added. From Lee Siegel, "Twitter Can't Save You," *New York Times Book Review*, February 4, 2011, 14–15. (Review of Evgeny Morozov's *The Net Delusion: The Dark Side of Internet Freedom* [New York: PublicAffairs, 2011]). http://www.nytimes .com/2011/02/06/books/review/Siegel-t.html?pagewanted=all&_r=0 (accessed October 12, 2016).

2. Italics added. "MIT 150: 150 Fascinating, Fun, Important, Interesting, Livesaving, Life-altering, Bizarre and Bold Ways that MIT Has Made a Difference," *Boston Globe*, Innovation supplement, May 15, 2011; see no. 12 in the list of 150, pages 16 and 18.

3. "Definitions of Digital Preservation," prepared by the ALCTS Preservation and Reformatting Section, Working Group on Defining Digital Preservation, ALA Annual Conference, Washington, D.C., June 24, 2007, http://www.ala.org/alcts/resources /preserv/defdigpres0408 (accessed December 10, 2016).

4. For a discussion about the importance of digital materials and the still-ongoing search for true digital preservation, see Sidney E. Berger, "Digital Preservation," in *Rare Books and Special Collections* (Chicago: Neal-Schuman/ALA, 2014), 390–396.

5. danah m. boyd and Nicole B. Ellison, "Social Network Sites: Definition, History, and Scholarship," *Journal of Computer-Mediated Communication* 13, no. 1 (2008): 211.

6. Craig Blaha, "Preserving Facebook Records: Subscriber Expectations and Behavior," *Preservation, Digital Technology & Culture* 42, no. 3 (2013): 115, 125.

7. Ibid., 126.

8. Ibid.

9. Digital artists are also concerned with preserving the look and feel of their early work. See Conor McGarrigle, "Preserving Born Digital Art: Lessons from Artists' Practice," *New Review of Information Networking* 20 (2015): 170–178, http://arrow.dit.ie /cgi/viewcontent.cgi?article=1034&context=aaschadpart (accessed January 29, 2017).

10. Derek L. Murphy, "Documenting Pocket Universes: New Approaches to Preserving Online Games," *Preservation, Digital Technology & Culture* 44, no. 4 (2015): 179.

11. See, for example, Jerome McDonough, "A Tangled Web: Metadata and Problems in Game Preservation," in *The Preservation of Complex Objects,* volume 3, *Gaming Environments and Virtual Worlds,* ed. Janet Delve et al. (London: JISC, 2013), 49–62.

12. Richard A. Bartle, "Archaeology Versus Anthropology: What Can Truly Be Preserved?" in *Preserving Complex Digital Objects,* ed. Janet Delve and David Anderson (London: Facet Publishing, 2014), 15.

13. Sabine Himmelsbach, "Net-based and Networked: Challenges for the Conservation of Digital Art," iPres 2016: 13th International Conference on Digital Preservation, Bern, Switzerland, Keynote talk no. 2, October 4, 2016.

14. Ibid.

15. Richard Rinehart and Jon Ippolito, *Re-Collection: Art, New Media, and Social Memory* (Cambridge, MA: MIT Press, 2014).

16. Ibid., 75.

17. Ibid., 83.

18. UNESCO, Charter on the Preservation of Digital Heritage, Paris: UNESCO, October 15, 2003, http://portal.unesco.org/en/ev.php-URL_ID=17721&URL_DO=DO _TOPIC&URL_SECTION=201.html (accessed December 21, 2016).

19. "The Memory of the World in the Digital Age: Digitization and Preservation: An International Conference on Permanent Access to Digital Documentary Heritage," Vancouver, BC, September 26–28, 2012.

20. The *UNESCO/PERSIST (Platform to Enhance the Sustainability of the Information Society Transglobally) Guidelines for the Selection of Digital Heritage for Long-Term Preservation* by the UNESCO/PERSIST Content Task Force, March 2016, grew out of the Vancouver 2012 conference (see note 19). See https://www.unesco.nl/sites/default /files/uploads/Comm_Info/persistcontentguidelinesfinal1march2016.pdf (accessed September 18, 2017).

21. UNESCO, Convention for the Safeguarding of the Intangible Cultural Heritage, Paris: UNESCO, October 17, 2003, http://portal.unesco.org/en/ev.php-URL_ID=17716 &URL_DO=DO_TOPIC&URL_SECTION=201.html (accessed December 21, 2016); and UNESCO, Charter on the Preservation of Digital Heritage.

22. Yola de Lusenet, "Tending the Garden or Harvesting the Fields: Digital Preservation and the UNESCO Charter on the Preservation of the Digital Heritage," *Library Trends* 56, no. 1 (Summer 2007): 174.

23. UNESCO, Convention for the Safeguarding of the Intangible Cultural Heritage, Article 2.

24. Abby Smith Rumsey, et al. *Sustainable Economies for a Digital Planet: Ensuring Long-Term Access to Digital Information.* Final Report of the Blue Ribbon Task Force on Sustainable Digital Preservation and Access. N.p., February 2010, at brtf.sdsc.edu (accessed November 28, 2017).

25. Samuel Jones and John Holden, *It's a Material World: Caring for the Public Realm* (London: Demos, 2008), 15.

26. Peter Johan Lor, and Johannes J. Britz, "An Ethical Perspective on Political-Economic Issues in the Long-Term Preservation of Digital Heritage," *Journal of the American Society for Information Science and Technology* 63, no. 11 (2012): 2153–2164.

27. Abby Smith Rumsey, *When We Are No More: How Digital Memory Is Shaping Our Future* (New York: Bloomsbury Press, 2016), 167.

28. Sabine Himmelsbach used this phrase in her keynote talk at iPres 2016, see Himmelsbach, "Net-based and Networked."

Chapter 9

1. John Ruskin, *The Storm-Cloud of the Nineteenth Century. Two Lectures Delivered at the London Institution February 4th and 11th, 1884* (Sunnyside, Orpington, Kent: George Allen, 1884), 36–37. Bound in *The Works of John Ruskin*, ed. Edward Tyas Cook and Alexander Wedderburn, vol. 34 (London: George Allen, 1908), 7–80.

2. Adele Berlin and Marc Zvi Brettler, eds., *The Jewish Study Bible*, Jewish Publication Society, Tanakh translation (Oxford and New York: Oxford University Press, 2004), 843 and 993.

3. Genesis 1:28; Berlin and Brettler, *The Jewish Study Bible*, 14.

4. Marcus Vitruvius Pollio (best known as Vitruvius), *De Architectura*, books 6.1.1–2 and 7.4.1–5 in any complete edition. It was written between 30 and 15 B.C.E., but not published until the Renaissance. See Morris Hicky Morgan, ed. and trans., *Vitruvius: The Ten Books on Architecture* (New York: Dover, 1960).

5. Gabriela Roxana Carone, "Plato and the Environment," *Environmental Ethics* 20, no. 2 (Summer 1998): 115.

6. For more about the Antiquities Act of 1906, see https://www.nps.gov/archeology /tools/laws/antact.htm (accessed July 10, 2016).

7. One organization that works with businesses, governments, and NGOs is First Peoples Worldwide, "first developed in 1997 by Cherokee social entrepreneur Rebecca Adamson, as a program of her non-profit First Nations Development Institute. In 2005, Rebecca and her daughter, Neva, founded First Peoples Worldwide as a full-fledged organization in its own right. [FPW focuses] on funding local development projects in Indigenous communities all over the world while creating bridges between our communities and corporations, governments, academics, NGOs and investors in their regions. [They] facilitate the use of traditional Indigenous knowledge in solving today's challenges, including climate change, food security, medicine, governance and sustainable development." See http://www.soanresources.com /indigenous-social-change.html (accessed September 20, 2017).

8. See United Nations Permanent Forum on Indigenous Issues, "Who Are Indigenous Peoples?" Factsheet, *Indigenous Peoples, Indigenous Voices*, http://www.un.org /esa/socdev/unpfii/documents/5session_factsheet1.pdf (accessed January 28, 2017).

9. Ruskin, *The Storm-Cloud of the Nineteenth Century*, 9.

10. Also referred to as the *devil-cloud* and *devil-wind*. Ibid., 67, in Ruskin's second lecture on the storm-cloud.

11. Ruskin, *The Storm-Cloud of the Nineteenth Century*, 3.

12. John Ruskin, "The Storm-Cloud of the Nineteenth Century, Lecture 1," reprinted in *The Norton Anthology of English Literature*, 8th ed., ed. M. H. Abrams. Available at https://www.wwnorton.com/college/english/nael/noa/pdf/27636_Vict _U08_Ruskin.pdf (accessed May 4, 2016).

13. Ibid., comment appears as footnote 1.

14. Brian J. Day, "The Moral Intuition of Ruskin's 'Storm-Cloud,'" *Studies in English Literature, 1500–1900* 45, no. 4 (Autumn 2005): 917–918.

15. Ibid., 918–919.

16. Ibid., 919.

17. Ruskin, *The Storm-Cloud of the Nineteenth Century*, 39.

18. Andrew Brennan, "Environmental Ethics," *Stanford Encyclopedia of Philosophy*, http://plato.stanford.edu/entries/ethics-environmental (accessed July 12, 2016).

19. John Ruskin, "The Lamp of Memory," in *The Seven Lamps of Architecture*, ed. Edward Tyas Cook and Alexander Wedderburn, vol. 8, section 18, 242.

20. Ibid., section 20, 244.

21. Ibid., 244.

22. In a letter that Ruskin wrote to Charles Eliot Norton on September 11, 1882, he pokes fun at Pisan architects and refers to the Leaning Tower. See *The Correspondence of John Ruskin and Charles Eliot Norton*, ed. John Lewis Bradley and Ian Ousby (Cambridge: Cambridge University Press, 1987), 450.

23. There is a good description in Wikipedia, "The Leaning Tower of Pisa," https://en.wikipedia.org/wiki/Leaning_Tower_of_Pisa (accessed July 10, 2016).

24. A daguerreotype of the church, possibly by Ruskin, was taken ca. 1846, before the reconstruction. See http://www.victorianweb.org/painting/ruskin/daguerrotypes/6.html (accessed May 22, 2016).

25. For a full discussion on this topic, see Michèle V. Cloonan, "The Moral Imperative to Preserve," *Library Trends* 55, no. 3 (Winter 2007): 746–755.

26. Rachel Carson, *Silent Spring* (Boston: Houghton Mifflin, 1962), table of contents, 16.

27. Quoted in K. D. [Kesharichand Dasharathasa] Gangrade, *Moral Lessons from Gandhi's Autobiography and Other Essays* (New Delhi: Concept Publishing Co., 2004), 211.

28. Rob Stewart, *Sharkwater: The Photographs* (Toronto: Key Porter Books, 2007).

29. Melanie L. Jackson, "The Principles of Preservation: The Influence of Viollet, Ruskin, and Morris on Historic Preservation," M.A. thesis, Oklahoma State University, May 2006, 95–101.

30. In this book I have been talking about monumental preservation and the many ways it manifests itself. One way—beyond my scope here—has to do with the preservation of human life. The vile air conditions clearly shortened people's lives.

31. Ken Taylor, Archer St. Clair, and Nora J. Mitchell, eds., *Conserving Cultural Landscapes: Challenges and New Directions* (New York: Routledge, 2014), xi–xii.

32. Roger J. H. King, "Environmental Ethics and the Built Environment," *Environmental Ethics* 22, no. 2 (Summer 2000): 115–131.

33. David Lowenthal, "Stewarding the Future," in *Preserving Our Heritage: Perspectives from Antiquity to the Digital Age,* ed. Michèle Valerie Cloonan (Chicago: Neal-Schuman, 2015), 623–636. Originally published in *CRM: The Journal of Heritage Stewardship* 2, no. 2 (Summer 2005): 1–17.

34. "Stewardship," *The American Heritage Dictionary,* 4th ed. (Boston: Houghton Mifflin, 2000).

35. Lowenthal cites Martin Rees who, writing in 2002, worried that bioterror or bio-error would lead to a million casualties in a single event within the next fifteen years. See Lowenthal, "Stewarding the Future," 632 and 636.

36. Jennifer Welchman, "The Virtues of Stewardship," *Environmental Ethics* 21, no. 4 (Winter 1999): 411–423.

37. Ibid., 414.

38. Janna Thompson, "Environment as Cultural Heritage," *Environmental Ethics* 22, no. 3 (Fall 2000): 241–258.

39. Emphasis in the original. See https://www.nps.gov/klgo/learn/education/class rooms/conservation-vs-preservation.htm (accessed September 20, 2017).

40. National Park Service, "Conservation vs. Preservation and the National Park Service," Lesson Plan, Updated April 14, 2015, https://www.nps.gov/klgo/learn/educa tion/classrooms/conservation-vs-preservation.htm (accessed on September 20, 2017).

41. David Baron, *The Beast in the Garden: A Modern Parable of Man and Nature* (New York: W. W. Norton, 2004), 237–238.

42. Marcus Hall, *Earth Repair: A Transatlantic History of Environmental Restoration* (Charlottesville: University of Virginia Press, 2005), 1–5.

43. See James Beck and Michael Daley, *Art Restoration: The Culture, the Business and the Scandal* (New York: W. W. Norton, 1995); Michael Kimmelman, "Finding God in a Double Foldout," *The New York Times,* December 8, 1991; and Andrew Wordsworth, "Have Italy's Art Restorers Cleaned up Their Act?" *The Independent,* June 20, 2000.

44. Pierluigi De Vecchi, et al., *The Sistine Chapel: A Glorious Restoration* (New York: Harry N. Abrams, 1994).

45. Salvador Muñoz Viñas, *Contemporary Theory of Conservation* (Amsterdam: Elsevier Butterworth-Heinemann, 2005), 205.

46. Ibid., 205–206.

47. For debates about cleaning *David,* see Alan Riding, "Spruced UP for a 500th Birthday; After His Bath, 'David' Has a Bit More Shine," *The New York Times,* "Arts," May 25, 2004, http://www.nytimes.com/2004/05/25/arts/spruced-up-for-a-500th -birthday-after-his-bath-david-has-a-bit-more-shine.html?_r=0 (accessed July 21,

2016). A new proposal for *David*'s preservation has recently been put forth since the 2016 earthquake in central Italy: to place the statue in an anti-seismic museum to protect *David*'s lower legs, which have microfractures.

48. David A. Scott, "Conservation and Authenticity: Interactions and Enquiries," *Studies in Conservation* 60, no. 5 (2015): 303.

49. Hall, *Earth Repair*, 2–3.

50. Ibid., 3.

51. Ibid., 5.

52. Page 1 of the digitized and printed version of George Perkins Marsh, *Man and Nature; or, Physical Geography as Modified by Human Action* (Memphis: General Books, 2012 [1864]).

53. Georges-Louis Leclerc, Comte de Buffon, wrote extensively about his views in *Histoire Naturelle*, a thirty-six-volume work. He is often cited for his wide-ranging ideas, which are said to include global warming. I mention him here as an example of someone who promoted the idea that people improved nature. See "George-Louis Leclerc, count de Buffon: French Naturalist," https://www.britannica.com/biography /Georges-Louis-Leclerc-comte-de-Buffon (accessed July 21, 2016); and University of California Museum of Paleontology, "Georges-Louis, Leclerc, Comte de Buffon (1707–1788)," http://www.ucmp.berkeley.edu/history/buffon2.html (accessed July 21, 2016).

54. Hall, *Earth Repair*, 14–15.

55. Marsh, *Man and Nature*, 210.

56. Hall, *Earth Repair*, 15.

57. Ibid., 16.

58. Brower wrote, "Between June 9, 1966, and April 16, 1967, the Sierra Club placed four full-page ads in *The New York Times* (some of them were repeated in many other newspapers and magazines) to carry forward the battle to keep dams out of the Grand Canyon. The first of the series was extraordinary in two ways. It was a split run—something the *Times* had never undertaken before in its daily paper. Half of the June 9 copies of the *Times* contained my relatively quiet open letter to Secretary Udall, asking him to help save the Grand Canyon and asking the public to speak up, too. It contained one coupon, addressed to the Sierra Club." From David Brower, "Grand Canyon Battle Ads," http://content.sierraclub.org/brower/grand-canyon -battle-ads (accessed July 10, 2016).

59. Bill Gammage, *The Biggest Estate on Earth: How Aborigines Made Australia* (Sydney: Allen & Unwin, 2011), xviii.

60. Ibid., 4.

61. Ibid., 304.

62. Thompson, "Environment as Cultural Heritage," 257.

63. Council of the European Union, *Conclusions on Cultural Heritage as a Strategic Resource for a Sustainable Europe*. Brussels: Council of the European Union, Education, Youth, Culture and Sport Council meeting, May 20, 2014.

64. Ibid., 1, conclusion 2.

65. Ibid., 1, conclusion 4.

66. Ibid., 2, conclusion 7.

Chapter 10

1. Michèle V. Cloonan, "The Paradox of Preservation," *Library Trends* 56, no. 1 (Summer 2007): 136.

2. Ibid.

3. Ibid.

4. *Merriam-Webster*, "Paradox," definition no. 3, http://www.merriam-webster.com/ (accessed December 1, 2016).

5. Cloonan, "The Paradox of Preservation," 138.

6. Jeff Spurr [now retired; then at Harvard University], personal communication, April 13, 2006, quoted in Cloonan, "The Paradox of Preservation," 141.

7. Shayee Khanaka [now retired; then at the University of California, Berkeley], personal communication, April 13, 2006, quoted in Cloonan, "The Paradox of Preservation," 141.

8. Paul Sturges, "Limits to Freedom of Expression? Considerations Arising from the Danish Cartoons Affair," *IFLA Journal* 32, no. 3 (2006): 188.

9. Paul Sturges, "Limits to Freedom of Expression? The Problem of Basphemy," *IFLA Journal* 41, no. 2 (2015): 112–119.

10. Christopher Mele, "Libraries Become Unexpected Sites of Hate Crimes," *The New York Times*, December 8, 2016, http://mobile.nytimes.com/2016/12/08/us /libraries-hate-crimes.html (accessed December 20, 2016).

11. *The New York Times*. Information can also be obtained from the website of the American Library Association, http://www.ala.org/.

12. The story is rather complicated; Mullar Omar's opinion changed and he has made contradictory claims about why he had the statues destroyed.

13. Christoph Machat, Michael Petzet, and John Ziesemer, eds., *Heritage at Risk: ICOMOS World Report, 2008–2010 on Monuments and Sites in Danger* (Berlin: Hendrik Bässler Verlag, 2010).

14. UNESCO, "Expert Working Group Releases Recommendations for Safeguarding Bamiyan" (Paris: UNESCO World Heritage Convention, April 27, 2011), whc.unesco .org (accessed December 1, 2016).

15. Robert Bevan, "Ruin or Rebuild? Conserving Heritage in an Age of Terrorism," *The Art Newspaper*, January 2017, section 2, 20–21.

16. Ibid., 20.

17. Cloonan, "The Paradox of Preservation," 145.

18. "Sisi Names Bomber of Cairo Coptic Christian Church," *Al Jazeera*, December 12, 2016.

19. Cloonan, "The Paradox of Preservation," 145.

Chapter 11

1. President Kennedy's quote in German appears on a brass plaque on the floor of the Mall of Berlin.

2. Renzo Piano, quoted in Peter Schneider, *Berlin Now: The City After the Wall*, trans. by Sophie Schlondorff (New York: Farrar, Straus and Giroux, 2014), p. 34.

3. Information that is not endnoted I have gleaned from visits to Berlin. A lot of information is available at memorial and construction sites or at visitor centers. I took copious notes and/or photographed the signs.

4. Svetlana Boym, *The Future of Nostalgia* (New York: Basic Books, 2001), 176. Later in the chapter she describes the city as "a stage set with many other temporary curtains—the wrapping of the Reichstag, the curtain on the Schloss Platz representing the facade of the destroyed building, and the curtain door of another destroyed masterpiece…, " 179.

5. *Encyclopaedia Britannica* 11th ed., "Berlin," https://en.wikisource.org/wiki/1911 _Encyclopædia_Britannica/Berlin (accessed November 30, 2016); and Kathryn W. Gunsch, "Seeing the World: Displaying Foreign Art in Berlin, 1898–1926," *Journal of Art Historiography* 12 (June 2015): 1–25. See https://arthistoriography.wordpress.com /12-jun-2015 (accessed December 2, 2016).

6. Ibid.

7. Gunsch, 2.

8. Robert M. Edsel, *The Monuments Men: Allied Heroes, Nazi Thieves, and the Greatest Treasure Hunt in History* (New York: Center Street, 2009), 10–16.

9. Text from the wall panel at the permanent exhibition of the Bode Museum, Berlin. The exhibit is called *Die Brände im Leitturm des Flakbunkers Friedrichshain im Mai 1945* (The Fires in the Main Tower of the Anti-Aircraft Bunker Friedrichshain in May 1945). Text from photograph taken by the author, April 2016.

10. Jürgen Scheunemann, *Top 10 Berlin* (London: DK Eyewitness Travel, 2014), 23.

11. Ibid., 10–11; see *Holocaust Encyclopedia*, "Holocaust Museum," at https://www .ushmm.org/learn/holocaust-encyclopedia (accessed December 2, 2016).

12. Deyan Sudjic, *Norman Foster: A Life in Architecture* (New York: Overlook Press, 2010), 198–204; 206–211.

13. I have used the terms *gays* and *homosexuals* because those are the words used on the information board across from the memorial.

14. Text transcribed from the information board near the memorial.

15. Peter Schneider, *Berlin Now: The City After the Wall*, trans. by Sophie Schlondorff (New York: Farrar, Straus and Giroux, 2014), 297–298.

16. Ibid., 299.

17. Richard Brody, "The Inadequacy of Berlin's "Memorial to the Murdered Jews of Europe," *The New Yorker*, July 12, 2012, http://www.newyorker.com/culture/richard -brody/the-inadequacy-of-berlins-memorial-to-the-murdered-jews-of-europe (accessed December 2, 2016).

18. Jewish Museum Berlin, "*Shalechet* (Fallen Leaves), Jewish Museum Berlin," art- work by Menashe Kadishman. April 2, 2013, https://fotoeins.com/2013/04/02 /shalechet-jewish-museum-berlin (accessed September 20, 2017).

19. Schneider, *Berlin Now*, 22.

20. Ibid., 35–37.

21. Scheunemann, *Top 10 Berlin*, 19.

22. Wikipedia [German language], "Hiroshimastrasse," https://de.wikipedia.org/wiki /Hiroshimastraße; accessed January 29, 2017.

23. Akira Kibi, "A Cool Breeze from Hiroshima," in *Revista Espaço Acadêmico* 51 (August 2005), http://www.espacoacademico.com.br/051rea.htm (accessed Decem- ber 3, 2016).

24. Schneider, *Berlin Now*, 55–56.

25. Various sources, including Schneider, *Berlin Now*; Jürgen Scheunemann, "Top 10 Plundered Artifacts," *Time Magazine*, n.d., http://content.time.com/time/specials/pack ages/article/0,28804,1883142_1883129_1883119,00.html (accessed December 3, 2016); and Ishaan Tharoor, "The Bust of Nefertiti: Remembering Ancient Egypt's Most

Famous Queen," *Time Magazine*, December 6, 2012, http://world.time.com/2012/12/06/the-bust-of-nefertiti-remembering-ancient-egypts-famous-queen (accessed December 3, 2016).

26. Rory McLean, "10 of the Best Museums in Berlin," *The Guardian*, August 17, 2011, https://www.theguardian.com/travel/2011/aug/17/top-10-museums-berlin-city-guides (accessed December 3, 2016).

Bibliography

"Afghanistan's Buddhas Rise Again." *The Atlantic*, June 10, 2015. https://www.the atlantic.com/international/archive/2015/06/3d-buddhas-afghanistan/395576; accessed November 11, 2016.

Agiesta, Jennifer. "Poll: Majority Sees Confederate Flag as Southern Pride Symbol, Not Racist." *CNN*, July 2, 2015. http://www.cnn.com/2015/07/02/politics/confeder ate-flag-poll-racism-southern-pride/index.html; accessed October 11, 2017.

Antiquities Act of 1906. National Park Service, U.S. Department of the Interior. Archeology Program. https://www.nps.gov/archeology/tools/laws/antact.htm; accessed July 10, 2016.

Art Institute of Chicago. "Archive of Architectural Photographer Richard Nickel Goes to Art Institute's Ryerson and Burnham Libraries." November 17, 2010. http://www.artic.edu/sites/default/files/press_pdf/Richard_Nickel_Archi.pdf; accessed February 28, 2016.

Báez, Fernando. *A Universal History of the Destruction of Books: From Ancient Sumer to Modern Iraq*. Trans. A. MacAdam. New York: Atlas & Co, 2004.

Baker, Aryn. "Afghanistan's Buddhas Can Be Rebuilt. But Should They?" *Time*, March 2, 2011. http://world.time.com/2011/03/02/afghanistans-buddhas-can-be -rebuilt-but-should-they; accessed October 12, 2015.

Baron, David. *The Beast in the Garden: A Modern Parable of Man and Nature*. New York: W. W. Norton, 2004.

Bartle, Richard A. "Archaeology Versus Anthropology: What Can Truly Be Preserved?" In *Preserving Complex Digital Objects*, ed. Janet Delve and David Anderson, 13–20. London: Facet Publishing, 2014.

Basbanes, Nicholas A. *A Gentle Madness: Bibliophiles, Bibliomanes, and the Eternal Passion for Books*. New York: Henry Holt & Co, 1995.

BBC News. "Life and Death in Syria: Five Years into War, What Is Left of the Country?" March 15, 2016. http://www.bbc.co.uk/news/resources/idt-841ebc3a-1be9-493b-8800-2c04890e8fc9; accessed August 18, 2016.

Beck, James, and Michael Daley. *Art Restoration: The Culture, the Business and the Scandal.* New York: W. W. Norton, 1995.

Berger, Sidney E. *The Dictionary of the Book: A Glossary for Book Collectors, Booksellers, Librarians, and Others.* Lanham, MD: Rowman & Littlefield, 2016.

Berger, Sidney E. *Rare Books and Special Collections.* Chicago: Neal-Schuman/ALA, 2014.

Berlin, Adele, and Marc Zvi Brettler, eds. *The Jewish Study Bible.* Jewish Publication Society, Tanakh translation. Oxford and New York: Oxford University Press, 2004.

Bevan, Robert. "Ruin or Rebuild? Conserving Heritage in an Age of Terrorism." *The Art Newspaper,* January 2017, section 2, 20–21.

Bey, Lee. "The Storied Mecca Apartments Live Again, Thanks to 'Life' Magazine Archives," June 21, 2012. http://www.wbez.org/blogs/lee-bey/2012-06/storied-mecca-apartments-live-again-thanks-life-magazine-archives-100301; accessed February 18, 2016.

Blaha, Craig. "Preserving Facebook Records: Subscriber Expectations and Behavior." *Preservation, Digital Technology & Culture* 42 (3) (2013): 115–128.

Blue Shield International. "International Committee of the Blue Shield (ICBS)." http://www.ancbs.org/cms/en/about-us/about-icbs; accessed August 28, 2016.

Bluestone, Daniel. *Buildings, Landscapes, and Memory: Case Studies in Historic Preservation.* New York: W. W. Norton, 2011.

Bluestone, Daniel. "Chicago's Mecca Flat Blues." In *Giving Preservation a History: Histories of Historic Preservation in the United States,* ed. Max Page and Randall Mason, 150 ff. New York and London: Routledge, 2004.

Bokova, Irina. "Discours de la Directrice générale de l'UNESCO Irina Bokova, à l'occasion de la Réunion de haut niveau pour la protection du patrimoine culturel syrien." August 29, 2013 (DG/2013/119) (French and English). http://www.unesco.org/new/fileadmin/MULTIMEDIA/HQ/ERI/pdf/ReunionPatrimoineSyrie.pdf; accessed July 3, 2015.

"Book of Kells for Exhibition in US." *Irish Times.* February 5, 1976, 7.

Boston Museum of Fine Arts. "Kunstkammer Gallery," n.d. http://www.mfa.org/collections/featured-galleries/kunstkammer-gallery?utm_source=google-grant&utm_medium=cpc&utm_content=gallery-kunstkammer_4&utm_campaign=mfa-brand&gclid=CNz72IXwss8CFQ9ahgodqCUCFw; accessed September 27, 2016.

boyd, danah m., and Nicole B. Ellison. "Social Network Sites: Definition, History, and Scholarship." *Journal of Computer-Mediated Communication* 13 (1) (2008): 210–230.

Boym, Svetlana. *The Future of Nostalgia*. New York: Basic Books, 2001.

Brody, Richard. "The Inadequacy of Berlin's Memorial to the Murdered Jews of Europe." *The New Yorker*, July 12, 2012. http://www.newyorker.com/culture/richard -brody/the-inadequacy-of-berlins-memorial-to-the-murdered-jews-of-europe; accessed December 2, 2016.

Brower, David. "Grand Canyon Battle Ads." http://content.sierraclub.org/brower /grand-canyon-battle-ads; accessed July 10, 2016.

Brown, Jean. "Jean Brown Papers, 1916–1995 (bulk 1958–1985)." Jean Brown Collection finding aid, Getty Research Institute. http://archives2.getty.edu:8082/xtf /view?docId=ead/890164/890164.xml;chunk.id=ref15;brand=default; accessed September 28, 2016.

Brown, Jonathan. *In the Shadow of Velázquez: A Life in Art History*. New Haven: Yale University Press, 2014.

Bush, Vannevar. "As We May Think." *Atlantic Monthly* 176 (1) (July 1945): 101–108.

Cafiero, Giorgio. "Hamas and Hezbollah Agree to Disagree on Syria." *Huffington Post* (January 31, 2014; updated April 2, 2014). http://www.huffingtonpost.com/giorgio -cafiero/hamas-and-hezbollah-agree_b_4698024.html; accessed August 15, 2016.

Cahan, Richard. *They All Fall Down: Richard Nickel's Struggle to Save America's Architecture*. New York: John Wiley & Sons, 1994.

Cahan, Richard, and Michael Williams. *Richard Nickel's Chicago: Photographs of a Lost City*. Chicago: CityFiles Press, 2008.

Carbone, Christopher. "Oldest Historian Group on Confederate Monuments: Preservation and 'Historical Context' Needed." *Fox News*, August 31, 2017. http://www .foxnews.com/us/2017/08/31/oldest-historians-group-on-confederate-monuments -preservation-and-historical-context-needed.html; accessed September 17, 2017.

Carone, Gabriela Roxana. "Plato and the Environment." *Environmental Ethics* 20 (2) (Summer 1998): 115–133.

Carson, Rachel. *Silent Spring*. Boston: Houghton Mifflin, 1962.

Carter, John. *Taste and Technique in Book Collecting: A Study of Recent Developments in Great Britain and the United States*. New York: R. R. Bowker, 1948.

Charny, Israel W. *How Can We Commit the Unthinkable? Genocide: The Human Cancer*. Boulder, CO: Westview Press, 1982.

Chicago Architecture Foundation. *Buildings of Chicago*. http://www.architecture.org /architecture-chicago/buildings-of-chicago; accessed February 28, 2016.

Christie's. "RELEASE: Safeguarding the Future of Russborough, Ireland—The Alfred Beit Foundation Will Offer A Small Group of Carefully Selected Old Master Paintings by Rubens, Adriaen van Ostade, David Teniers the Younger & Francesco Guardi." http://www.christies.com/about-us/press-archive/details?PressReleaseID=7887& lid=1&mob-is-app=false; accessed September 20, 2017.

Chung, Jeanie. "In the Fray." Interview with Matthew Barber. *The University of Chicago Magazine* 108 (2) (Winter 2016). http://mag.uchicago.edu/law-policy-society /fray; accessed January 25, 2017.

Clark, Kenneth. *One Hundred Details from Pictures in the National Gallery.* London: The National Gallery, 1938.

Cloonan, Michèle V. Holograph diary, August 17–28, 1991, possession of the author.

Cloonan, Michèle V. "Monumental Preservation: A Call to Action." *American Libraries* 35 (8) (September 2004): 34–38.

Cloonan, Michèle V. "The Moral Imperative to Preserve." *Library Trends* 55 (3) (Winter 2007): 746–755.

Cloonan, Michèle V. "The Paradox of Preservation." *Library Trends* 56 (1) (Summer 2007): 133–147.

Cloonan, Michèle Valerie, ed. *Preserving Our Heritage: Perspectives from Antiquity to the Digital Age.* Chicago: Neal-Schuman; London: Facet, 2015.

Cloonan, Michèle V., Allison Cuneo, and Father Columba Stewart. "Update on Syria." *Preservation, Digital Technology & Culture* 45 (2) (2016). DOI 10.1515/pdtc-2016-0015. http://www.hmml.org/about-us.html; accessed August 28, 2016.

Cocks, Anna Somers. "Italy's Museums Honour Archaeologist Murdered by Isil: Flags Were Flown at Half-mast around the Country and Heritage Experts Shared Their Outrage." *The Art Newspaper,* August 21, 2015. http://theartnewspaper.com/news/news /italy-s-museums-honour-archaeologist-murdered-by-isil; accessed August 2016.

Conway, Paul. "Preserving Imperfection: Assessing the Incidence of Digital Imaging Error in HathiTrust." *Preservation, Digital Technology & Culture* 42 (1) (2013): 17–30. https://deepblue.lib.umich.edu/bitstream/handle/2027.42/99522/J23 Conway Preserving Imperfection 2013.pdf?sequence=1&isAllowed=y; accessed January 28, 2017.

Cooper, John. *Raphael Lemkin and the Struggle for the Genocide Convention.* London: Palgrave Macmillan, 2008.

Cornell University Law School. "Federal Rules of Evidence: Article X. Contents of Writings, Recordings, and Photographs." https://www.law.cornell.edu/rules/fre /article_X; accessed October 31, 2016.

Council of the European Union. *Conclusions on Cultural Heritage as a Strategic Resource for a Sustainable Europe.* Brussels: Council of the European Union. Education, Youth,

Culture and Sport Council meeting, Brussels, May 20, 2014. https://www.consilium .europa.eu/uedocs/cms_data/docs/pressdata/en/educ/142705.pdf; accessed January 28, 2017.

Coyle, Karen. "Mass Digitization of Books." *Journal of Academic Librarianship* 32 (6) (November 2006): 641–645.

Craig, Barbara L. "Selected Themes in the Literature on Memory and Their Perti- nence to Archives." *American Archivist* 65 (2) (Fall/Winter 2002): 276–289.

Cuno, James. *Who Owns Antiquity? Museums and the Battle over Our Ancient Heritage.* Princeton: Princeton University Press, 2008.

Davis, Craig S. *The Middle East for Dummies.* Hoboken, NJ: Wiley Publishing, 2003.

Davy, Steven. "They Were Destroyed by the Taliban. But Now the Giant Buddha Statues of Bamiyan Have Returned with 3-D Light Projection." *PRI's The World,* June 11, 2015 http://www.pri.org/stories/2015-06-11/they-were-destroyed-taliban-now -giant-buddha-statues-bamiyan-have-returned-3-d; accessed October 12, 2015.

Dawisha, Adeed I. *Syria and the Lebanese Crisis.* New York: St. Martin's Press, 1980.

Day, Brian J. "The Moral Intuition of Ruskin's 'Storm-Cloud.'" *Studies in English Lit- erature, 1500–1900* 45 (4) (Autumn 2005): 917–933.

DCMS. "Hague Convention 1954." Department for Culture, Media and Sport. http:// old.culture.gov.uk/what_we_do/cultural_property/6630.aspx; accessed July 18, 2015.

DeHart, Robert. "Cultural Memory." In *Encyclopedia of Library and Information Sci- ences.* 3rd ed., ed. Marcia J. Bates and Mary Niles Maack, 1363. Boca Raton, FL: CRC Press, 2010.

De Lusenet, Yola. "Tending the Garden or Harvesting the Fields: Digital Preservation and the UNESCO Charter on the Preservation of the Digital Heritage." *Library Trends* 56 (1) (Summer 2007): 164–182.

de Paor, Liam. "The Book of Kells." *Irish Times,* February 24, 1976, 8.

de Paor, Liam. "Irish Treasures for America." *Irish Times,* July 8, 1977, 8.

De Vecchi, Pierluigi, et al. *The Sistine Chapel: A Glorious Restoration.* New York: Harry N. Abrams, 1994.

de Waal, Thomas. "The G-Word: The Armenian Massacre and the Politics of Geno- cide." *Foreign Affairs* 94 (1) (January/February 2015): 136–148.

Dibdin, Thomas Frognall. *The Bibliomania; Or, Book-Madness; Containing Some Account of the History, Symptoms, and Cure of This Fatal Disease. In an Epistle Addressed to Richard Heber.* London: Longman, Hurst, Rees, and Orme, 1809, and many reprints.

Donne, John. *Devotions upon Emergent Occasions,* "Meditation XVII," 1624. Frequently reprinted. See http://triggs.djvu.org/djvu-editions.com/DONNE/DEVO TIONS/Download.pdf, 31; accessed January 25, 2017.

Duguid, Paul. "Inheritance and Loss: A Brief Survey of Google Books." *First Monday* 12 (8) (August 6, 2007). http://firstmonday.org/ojs/index.php/fm/article/view/1972 /1847; accessed October 27, 2016.

Dunant, Henry. *Memory of Solferino.* Geneva: International Committee of the Red Cross, 1862; repr. 1986.

Dupré, Judith. *Monuments: America's History in Art and Memory.* New York: Random House, 2007.

Edsel, Robert M. *The Monuments Men: Allied Heroes, Nazi Thieves, and the Greatest Treasure Hunt in History.* New York: Center Street, 2009.

"Elena Baturina." Profile. *Forbes,* n.d. http://www.forbes.com/profile/elena-baturina/; accessed December 1, 2016.

Encyclopaedia Britannica, 11th ed. "Berlin." https://en.wikisource.org/wiki/1911_ Encyclopædia_Britannica/Berlin; accessed November 30, 2016.

Ephemera Society of America. "Ephemera." n.d. http://www.ephemerasociety.org /def.html; accessed September 28, 2016.

Falser, Michael. "The Bamiyan Buddhas, Performative Iconoclasm and the 'Image' of Heritage." In *The Image of Heritage: Changing Perception, Permanent Responsibilities,* ed. Simone Giometti and Andrzej Tomaszewski, 157–169. Proceedings of the International Conference of the ICOMOS International Scientific Committee for the Theory and the Philosophy of Conservation and Restoration, 6–8 March 2009. Florence, Italy: ICOMOS 2011.

Fearrington, Florence. *Rooms of Wonder: From Wunderkammer to Museum, 1599–1899: An Exhibition at the Grolier Club, 5 December 2012–2 February 2013.* New York: The Grolier Club, 2012.

Federal Rules of Evidence. Article X. "Contents of Writings, Recordings, and Photographs," Rule 1003, "Admissibility of Duplicates." https://www.law.cornell.edu/rules /fre/article_X; accessed October 31, 2016.

Fildis, Ayse Tekdal. "The Troubles in Syria: Spawned by French Divide and Rule." *Middle East Policy Council* 18 (4) (Winter 2011): 129–139; http://www.mepc.org /troubles-syria-spawned-french-divide-and-rule; accessed August 18, 2016.

First Peoples Worldwide. "About Us." http://www.soanresources.com/indigenous -social-change.html; accessed September 20, 2017.

Fisher, Max. "The One Map that Shows Why Syria Is So Complicated." *The Washington Post,* August 27, 2013. https://www.washingtonpost.com/news/worldviews/wp

/2013/08/27/the-one-map-that-shows-why-syria-is-so-complicated; accessed August 19, 2016.

Forest, Benjamin, and Juliet Johnson. "Unraveling the Threads of History: Soviet-Era Monuments and Post-Soviet National Identity in Moscow." *Annals of the Association of American Geographers* 92 (3) (September 2002): 524–547.

Freud, Sigmund. "A Disturbance of Memory on the Acropolis: An Open Letter to Romain Rolland on the Occasion of His Seventieth Birthday." The text was dated by Freud, January 1936. http://www.scribd.com/doc/6815599/A-Disturbance-of-Memory -on-the-Acropolis; accessed July 31, 2015. This text was no longer accessible in October 2017, but see https://s3.amazonaws.com/freud2lacan.docs/A_Disturbance_of _Memory.pdf.

Frieze, Donna-Lee, ed. *Journal of Genocide Research* 15 (3) (2013): 247–377.

Gammage, Bill. *The Biggest Estate on Earth: How Aborigines Made Australia.* Sydney: Allen & Unwin, 2011.

Gangrade, K. D. [Kesharichand Dasharathasa]. *Moral Lessons from Gandhi's Autobiography and Other Essays.* New Delhi: Concept Publishing Co, 2004.

The Getty Institute. Getty Research Institute (GRI) finding aid. http://archives2 .getty.edu:8082/xtf/view?docId=ead/890164/890164.xml;query=;brand=default; accessed September 26, 2016.

Giermann, Holly. "The Destruction of a Classic: Time-Lapse Captures Demolition of Chicago's Prentice Women's Hospital." February 4, 2015. http://www.archdaily.com /595078/the-destruction-of-a-classic-time-lapse-captures-demolition-of-chicago-s -prentice-women-s-ospital; accessed February 4, 2016.

The Guardian. "Beheaded Syrian Scholar Refused to Lead Isis to Hidden Palmyra Antiquities." https://www.theguardian.com/world/2015/aug/18/isis-beheads-archaeologist -syria; accessed August 2016.

Gunsch, Kathryn W. "Seeing the World: Displaying Foreign Art in Berlin, 1898–1926." *Journal of Art Historiography* 12 (June 2015): 1–25. https://arthistoriography .wordpress.com/12-jun-2015; accessed December 2, 2016.

Gunter, Michael M. *Out of Nowhere: The Kurds of Syria in Peace and War.* London: Hurst, 2015.

Halasa, Malu, Zaher Omareen, and Nawara Mahfoud, eds. *Syria Speaks: Art and Culture from the Frontline.* London: Saqi Books, 2014.

Hall, Marcus. *Earth Repair: A Transatlantic History of Environmental Restoration.* Charlottesville: University of Virginia Press, 2005.

Hall, Marcus, ed. *Restoration and History: The Search for a Usable Environmental Past.* London: Routledge, 2010.

"Henry Dunant—Biographical." Nobelprize.org. http://www.nobelprize.org/nobel
_prizes/peace/laureates/1901/dunant-bio.html; accessed June 30, 2015.

Heritage for Peace. "Syria's Culture and Heritage." http://www.heritageforpeace.org
/syria-culture-and-heritage; accessed July 11, 2015.

Herscher, Andrew. *Violence Taking Place: The Architecture of the Kosovo Conflict*. Palo
Alto, CA: Stanford University Press, 2010.

Herscher, Andrew, and András Riedlmayer. *Destruction of Cultural Heritage in Kosovo,
1998–1999: A Post-war Survey* (Expert Report for the International Criminal Tribunal
for the Former Yugoslavia). The Miloševic Trial Public Archive. *HRP Bard*. http://hague
.bard.edu/reports/hr_riedlmayer-28feb2002.pdf; accessed November 30, 2017.

Himmelsbach, Sabine. "Net-based and Networked: Challenges for the Conservation
of Digital Art," iPres 2016: 13th International Conference on Digital Preservation,
Bern, Switzerland, Keynote talk no. 2, October 4, 2016. Abstract available at http://
www.ipres2016.ch/frontend/index.php?page_id=1166; accessed January 29, 2017.

Holmes, Carolyn E. "Should Confederate Monuments Come Down? Here's What
South Africa Did after Apartheid." *The Washington Post*, August 29, 2017. https://
www.washingtonpost.com/news/monkey-cage/wp/2017/08/29/should-confederate
-monuments-come-down-heres-what-south-africa-did-after-apartheid/?utm_term
=.80e6105a9ce2; accessed September 17, 2017.

Holocaust Museum. "The Enabling Act." March 24, 1933. https://www.ushmm.org
/wlc/en/article.php?ModuleId=10007892; accessed January 29, 2017.

Hosking, Geoffrey A. "Memory in a Totalitarian Society: The Case of the Soviet
Union." In *Memory: History, Culture and the Mind*, ed. Thomas Butler, 115–130.
Oxford: Basil Blackwell, 1989.

Hosking, Geoffrey A., and Yitzhak M. Brudny. *Reinventing Russia: Russian Nationalism and the Soviet State, 1953–1991*. Cambridge, MA: Harvard University Press,
2000.

ICBS (International Committee of the Blue Shield). The Seoul Declaration on the
Protection of Cultural Heritage in Emergency Situations (December 9, 2011). http://
icom.museum/uploads/media/111210_ICBS_seoul_declaration_final.pdf; accessed
August 28, 2016.

ICOMOS. "ICOMOS Supports the Monuments Women and Men of Syria and Iraq."
http://www.icomos.org/en/178-english-categories/news/4084-icomos-supports-the
-monuments-women-and-men-of-syria-and-iraq; accessed October 12, 2015.

ICOMOS. International Charter for the Conservation and Restoration of Monuments and Sites (The Venice Charter 1964). https://www.icomos.org/charters
/venice_e.pdf; accessed January 28, 2017.

Internet Archive. "Scanning Services: Digitizing Print Collections with the Internet Archive." https://archive.org/scanning; accessed October 30, 2016.

Irvin-Erickson, Douglas. "Genocide, The 'Family of Mind' and the Romantic Signature of Raphael Lemkin." *Journal of Genocide Research* 15 (3) (2013): 273–296.

"Items Left at the Wall—The Virtual Collection." Vietnam Veterans Memorial Fund. http://www.vvmf.org/items/; accessed September 9, 2016.

Jackson, Holbrook. *The Anatomy of Bibliomania.* London: Soncino Press, 1930. Later reprinted.

Jackson, Melanie L. "The Principles of Preservation: The Influence of Viollet, Ruskin, and Morris on Historic Preservation." M.A. thesis, Oklahoma State University, May 2006.

Jacobs, Steven Leonard, ed. *Lemkin on Genocide.* Lanham, MD: Lexington Books, 2012.

Jampol, Justinian A. "Smashing Lenin Won't Save Ukraine." *New York Times*, Opinion and Editorial, March 4, 2014. http://www.nytimes.com/2014/03/04/opinion/smashing-lenin-wont-save-ukraine.html; accessed June 21, 2015.

Jefferson, Thomas. [Letter] "From Thomas Jefferson to Ebenezer Hazard, 18 February 1791." https://founders.archives.gov/documents/Jefferson/01-19-02-0059; accessed September 18, 2017.

Jefferson, Thomas. "Thomas Jefferson. Jefferson's Library." https://www.loc.gov/exhibits/jefferson/jefflib.html; accessed September 26, 2016.

Jewish Museum Berlin. "Shalechet (Fallen Leaves), Jewish Museum Berlin." Artwork by Menashe Kadishman. April 2, 2013. https://fotoeins.com/2013/04/02/shalechet-jewish-museum-berlin; accessed December 3, 2016.

"John Vinci: The Modern Preservationist." *Architects + Artisans: Thoughtful Design for a Sustainable World.* October 8, 2014. http://architectsandartisans.com/2014/10/john-vinci-the-modern-preservationist/; accessed February 9, 2016.

Jones, Samuel, and John Holden. *It's a Material World: Caring for the Public Realm.* London: Demos, 2008.

Khanaka, Shayee. Personal communication, April 13, 2006.

Kibi, Akira. "A Cool Breeze from Hiroshima." *Revista Espaço Acadêmico* 51 (August 2005). http://www.espacoacademico.com.br/051rea.htm; accessed December 3, 2016.

Kimmelman, Michael. "Finding God in a Double Foldout." *New York Times*, December 8, 1991. http://www.nytimes.com/1991/12/08/books/finding-god-in-a-double-foldout.html; accessed January 28, 2017.

King, Roger J. H. "Environmental Ethics and the Built Environment." *Environmental Ethics* 22 (2) (Summer 2000): 115–131.

Kishkovsky, Sophia. "Putin Dresses-down Culture Department over State of Monuments." *The Art Newspaper*, February 17, 2016. http://theartnewspaper.com/news /conservation/putin-s-dress-down-for-culture-department-over-state-of-monuments; accessed September 9, 2016.

Knuth, Rebecca. *Libricide: The Regime-Sponsored Destruction of Books and Libraries in the Twentieth Century*. Westport, CT: Praeger, 2003.

Leigh, Karen. "Hope for Palmyra's Future." *The Wall Street Journal*, April 5, 2016, Arts in Review, D5.

Lemkin, Raphael. *Axis Rule in Occupied Europe: Laws of Occupation, Analysis of Government, Proposals for Redress*. 2nd ed. Clark, NJ: The Lawbook Exchange, 2008. First published in 1944 in Washington, D.C., by the Carnegie Endowment for International Peace, Division of International Law. This "new" edition is a reprint of the 1944 edition with another, new, introduction.

Lemkin, Raphael. "Genocide—A Modern Crime." *Free World* 4 (April 1945): 39–43. http://www.preventgenocide.org/lemkin/freeworld1945.htm; accessed January 25, 2017.

Lieber, Francis. *General Orders No. 100: The Lieber Code*. "Instructions for the Government of Armies of the United States in the Field." Prepared by F. Lieber, promulgated as General Orders No. 100 by President Lincoln, April 24, 1863. The Avalon Project, Documents in Law, History and Diplomacy. Avalon.law.yale.edu/19th_century /lieber.asp#sec2; accessed June 28, 2015.

Lipe, William D. "A Conservation Model for American Archaeology." *The Kiva* 39 (3–4) (1974): 213–245.

Lor, Peter Johan, and Johannes J. Britz. "An Ethical Perspective on Political-Economic Issues in the Long-Term Preservation of Digital Heritage." *Journal of the American Society for Information Science and Technology* 63 (11) (2012): 2153–2164.

Lord, Albert B. *Singer of Tales*. Cambridge, MA: Harvard University Press, 1960. Reissued in an anniversary edition in 2002.

Lowenthal, David. *The Past Is a Foreign Country—Revisited*. Cambridge: Cambridge University Press, 2015.

Lowenthal, David. "Stewarding the Future." In *Preserving Our Heritage: Perspectives from Antiquity to the Digital Age*, ed. Michèle Valerie Cloonan, 623–636. Chicago: Neal-Schuman, 2015. Originally published in *CRM: The Journal of Heritage Stewardship* 2 (2) (Summer 2005): 1–17.

MacArthur Foundation. "Carl Haber: Audio Preservationist," *MacArthur Fellows Program,* September 25, 2013. https://www.macfound.org/fellows/class/2013; accessed September 17, 2017.

Machat, Christoph, Michael Petzet, and John Ziesemer, eds. *Heritage at Risk: ICOMOS World Report, 2008–2010 on Monuments and Sites in Danger.* Berlin: Hendrik Bässler Verlag, 2010.

Mako, Shamiran. "Cultural Genocide and Key International Instruments: Framing the Indigenous Experience." *International Journal on Minority and Group Rights* 19 (2012): 175–194.

Manaktala, Meghna. "Defining Genocide." *Peace Review: A Journal of Social Justice* 24 (2) (2012): 179–186.

Margottini, Claudio, ed. *After the Destruction of Giant Buddha Statues in Bamiyan (Afghanistan) in 2001: A UNESCO's Emergency Activity for the Recovery and Rehabilitation of Cliff and Niches.* Berlin: Springer, 2014.

Marsh, George Perkins. *Man and Nature; or, Physical Geography as Modified by Human Action.* Memphis: General Books, 2012. (Original ed. New York: Scribner and Co., 1864.)

McCarthy, Muriel. "Travelling Treasures." Letters to the Editor. *Irish Times*, July 13, 1977, 9.

McDonough, Jerome. "A Tangled Web: Metadata and Problems in Game Preservation." In *The Preservation of Complex Objects*, volume 3, *Gaming Environments and Virtual Worlds*, ed. Janet Delve et al., 49–62. London: JISC, 2013.

McGarrigle, Conor. "Preserving Born Digital Art: Lessons from Artists' Practice." *New Review of Information Networking* 20 (2015): 170–178. http://arrow.dit.ie/cgi/viewcontent.cgi?article=1034&context=aaschadpart; accessed January 29, 2017.

McGimsey, Charles Robert. *Public Archaeology.* New York: Seminar Press, 1972.

McLean, Rory. "10 of the Best Museums in Berlin." *The Guardian,* August 17, 2011. https://www.theguardian.com/travel/2011/aug/17/top-10-museums-berlin-city-guides; accessed December 3, 2016.

McMillan, M. E. *From the First World War to the Arab Spring: What's Really Going on in the Middle East?* New York: Palgrave Macmillan, 2016.

Mele, Christopher. "Libraries Become Unexpected Sites of Hate Crimes." *The New York Times,* December 8, 2016. mobile.nytimes.com/2016/12/08/us/libraries-hate-crimes.html; accessed December 20, 2016.

"MIT 150: 150 Fascinating, Fun, Important, Interesting, Livesaving, Life-altering, Bizarre and Bold Ways that MIT Has Made a Difference." *Boston Globe* Innovation supplement, May 15, 2011, no. 12, pp. 16, 18.

Mitchell, G. Frank, et al. *Treasures of Early Irish Art, 1500 BC–1500 AD.* New York: Metropolitan Museum of Art, 1977.

Mogg, William Rees. *How to Buy Rare Books: A Practical Guide to the Antiquarian Book Market*. Oxford: Phaidon, Christie's, 1985.

Monnens, Devin, Zach Vowell, Judd Ethan Ruggill, Ken S. McAllister, and Andrew Armstrong. "Before It's Too Late: A Digital Game Preservation White Paper." *American Journal of Play* 2 (2) (Fall 2009): 139–166.

Morgan, Morris Hicky, ed. and trans. *Vitruvius: The Ten Books on Architecture*. New York: Dover, 1960.

Morris, Sarah Rogers. "Richard Nickel's Photography: Preserving Ornament in Architecture." *Future Anterior: Journal of Historic Preservation, History, Theory, and Criticism* 10 (2) (Winter 2013): 66–80.

Morrison, Hugh. *Louis Sullivan: Prophet of Modern Architecture*. New York: Museum of Modern Art and W. W. Norton, 1935.

Muensterberger, Werner. *Collecting: An Unruly Passion: Some Psychological Perspectives*. Princeton: Princeton University Press, 1994.

Munby, A. N. L. *Phillipps Studies*. 5 vols. Cambridge: Cambridge University Press, 1951–1960.

Munby, A. N. L. *Portrait of an Obsession: The Life of Sir Thomas Phillipps, The World's Greatest Book Collector*. Adapted by Nicolas Barker. New York: G. P. Putnam's Sons; London: Constable, 1967.

Muñoz Viñas, Salvador. *Contemporary Theory of Conservation*. Amsterdam: Elsevier Butterworth-Heinemann, 2005.

Murphy, Derek. "Documenting Pocket Universes: New Approaches to Preserving Online Games." *Preservation, Digital Technology & Culture* 44 (4) (2015): 179–185.

Murray, Makenna. "Building Digital Preservation Capacity in the Middle East through Training: Discussing Training in Lebanon and the Future of Project Anqa with CyArk Field Manager, Ross Davison." July 28, 2016. http://www.cyark.org/news/building-digital-preservation-capacity-in-the-middle-east-through-training; accessed August 28, 2016.

Musteata, Sergiu, and Stefan Caliniuc, eds. *Current Trends in Archaeological Heritage Preservation: National and International Perspectives*. Proceedings of the International Conference, Iaşi, Romania, November 6–10, 2013. Oxford, UK: Archaeopress, 2015.

National Library of Medicine. *NLM Microfilming Project, Statement of Work*, revised May 17, 2001. https://www.nlm.nih.gov/psd/pres/contracts/filmsowfy01.txt; accessed October 30, 2016.

National Park Service. "Conservation vs. Preservation and the National Park Service." Lesson Plan. Updated April 14, 2015. https://www.nps.gov/klgo/learn/education/classrooms/conservation-vs-preservation.htm; accessed on September 20, 2017.

National Park Service. "Historic American Buildings Survey (HABS)," Heritage Documentation Programs. www.nps.gov/hdp/habs; accessed February 8, 2016.

"News: United Kingdom." *The Art Newspaper*, no. 269, June 2015, 10.

New York Times. "Jean Brown 82, Avid Collector of Dada, Surrealism and Fluxus." Obituaries, May 4, 1994. http://www.nytimes.com/1994/05/04/obituaries/jean -brown-82-avid-collector-of-dada-surrealism-and-fluxus.html; accessed November 20, 2016.

Nickel, Richard, and Aaron Siskind, with John Vinci and Ward Miller. *The Complete Architecture of Adler & Sullivan*. Chicago: Richard Nickel Committee, 2010.

"1973 Law of the Union of Soviet Socialist Republics: On the Protection and Use of Historic and Cultural Monuments." "Special Issue on the Preservation of Soviet Heritage," ed. Jean-Louis Cohen and Barry Bergdoll. *Future Anterior: Journal of Historic Preservation, History, Theory, and Criticism* 5 (1) (Summer 2008): 74–80.

Nora, Pierre. *Realms of Memory (Les Lieux de Mémoire)*. https://en.wikipedia.org/wiki /Les_Lieux_de_Mémoire; accessed October 2016.

Nye, Joseph. *Soft Power: The Means to Success in World Politics*. New York: Public Affairs, 2004.

Page, Max, and Randall Mason, eds. "Introduction: Rethinking the Roots of the Historic Preservation Movement." In *Giving Preservation a History: Histories of Historic Preservation in the United States*. New York and London: Routledge, 2004, 3–16. https:// books.google.com/books?id=owgcDRTKLxUC&pg=PA3&source=gbs_toc_r&cad=3 - v =onepage&q&f=fals; accessed January 28, 2017.

Parcak, Sarah. "Archaeological Looting in Egypt: A Geospatial View (Case Studies from Saqqara, Lisht, and el Hibeh)." *Near Eastern Archaeology* 78 (3) (2015): 196–203.

Payton, Bre. "Poll: Most Americans Don't Want Confederate Statues Torn Down." *The Federalist*, August 17, 2017. http://thefederalist.com/2017/08/17/poll-overwhelming -majority-americans-want-keep-confederate-statues; accessed September 17, 2017.

Pearce, Susan M., Rosemary Flanders, Mark Hall, Fiona Morton, Alexandra Bounia, and Paul Martin, eds. *The Collector's Voice: Critical Readings in the Practice of Collecting*. 4 vols. Aldershot, England: Ashgate, 2002.

Pearlstein, Ellen. "Conservation and Preservation of Museum Objects." In *Encyclopedia of Library and Information Sciences*. 3rd ed., ed. Marcia J. Bates and Mary Niles Maack, 1269–1276. Boca Raton, FL: Taylor & Francis, 2010.

Premiyak, Liza. "A Wilderness in the City: How Diller Scofidio + Renfro's Zaryadye Park Could Help Fix Moscow." ArchDaily, February 15, 2015, http://www.archdaily .com/598907/a-wilderness-in-the-city-how-diller-scofidio-renfro-s-zaryadye-park -could-help-fix-moscow; accessed September 18, 2017.

Prescott, Kurt. "Seven Things You Should Know about ASOR's Syrian Heritage Initiative," n.d. American School of Oriental Research. http://asorblog.org/2014/09/10/6-things-you-need-to-know-about-asors-syrian-heritage-initiative; accessed August 24, 2016.

Price, Nicholas Stanley, M. Kirby Talley Jr., and Alessandra Melucco Vaccaro. *Historical and Philosophical Issues in the Conservation of Cultural Heritage*. Los Angeles: The Getty Conservation Institute, 1996.

Puls, Keith E., ed. *Law of War Handbook*. JA 423. Charlottesville, VA: The Judge Advocate General's Legal Center and School, 2005. https://www.loc.gov/rr/frd/Military_Law/pdf/law-war-handbook-2004.pdf; accessed September 20, 2017.

Radley, Alan. "Artifacts, Memory and a Sense of the Past." In *Collective Remembering*, ed. David Middleton and Derek Edwards, 46–59. London: Sage, 1990.

Rhodes, Barbara J., and William W. Streeter. *Before Photocopying: The Art and History of Mechanical Photocopying, 1780–1938*. New Castle, DE: Oak Knoll, 1999.

Riding, Alan. "Spruced UP for a 500th Birthday; After His Bath, 'David' Has a Bit More Shine." *The New York Times*, May 25, 2004, Arts. http://www.nytimes.com/2004/05/25/arts/spruced-up-for-a-500th-birthday-after-his-bath-david-has-a-bit-more-shine.html?_r=0; accessed July 21, 2016.

Riedlmayer, András J. "Crimes of War, Crimes of Peace: Destruction of Libraries During and After the Balkan Wars of the 1990s." *Library Trends* 56 (1) (Summer 2007): 107–132.

Riegl, Alois. "The Modern Cult of Monuments: Its Essence and Its Development." Trans. Karin Bruckner with Karen Williams. In *Historical and Philosophical Issues in the Conservation of Cultural Heritage*, ed. Nicholas Stanley Price, M. Kirby Talley, Jr., and Alessandra Melucco Vaccaro, 69–83. Los Angeles: The Getty Conservation Institute, 1996.

Rinehart, Richard, and Jon Ippolito. *Re-Collection: Art, New Media, and Social Memory*. Cambridge, MA: MIT Press, 2014.

Rosen, Larry. "Is It Live or Is It Memorex? Viewing the World through a Camera Lens." *Psychology Today*, July 18, 2013. https://www.psychologytoday.com/blog/rewired-the-psychology-technology/201307/is-it-live-or-is-it-memorex; accessed October 20, 2016.

Rumsey, Abby Smith. *When We Are No More: How Digital Memory Is Shaping Our Future*. New York: Bloomsbury Press, 2016.

Rumsey, Abby Smith, et al. *Sustainable Economies for a Digital Planet: Ensuring Long-Term Access to Digital Information*. Final Report of the Blue Ribbon Task Force on Sustainable Digital Preservation and Access. N.p., February 2010. brtf.sdsc.edu; accessed November 28, 2017.

Rushby, Kevin. "Remembering Syria's Historic Silk Road in Aleppo." *The Guardian*, October 5, 2012. https://www.theguardian.com/travel/2012/oct/05/aleppo-souk -syria-destroyed-war; accessed August 17, 2016.

Ruskin, John. *The Correspondence of John Ruskin and Charles Eliot Norton*, ed. John Lewis Bradley and Ian Ousby. Cambridge: Cambridge University Press, 1987.

Ruskin, John. "The Lamp of Memory." In *The Seven Lamps of Architecture*, ed. Edward Tyas Cook and Alexander Wedderburn, vol. 8, 221 ff. London: George Allen, 1905.

Ruskin, John. "The Storm-Cloud of the Nineteenth Century, Lecture 1," reprinted in *The Norton Anthology of English Literature*, 8th ed., ed. M. H. Abrams. https://www .wwnorton.com/college/english/nael/noa/pdf/27636_Vict_U08_Ruskin.pdf; accessed May 4, 2016.

Ruskin, John. *The Storm-Cloud of the Nineteenth Century. Two Lectures Delivered at the London Institution February 4th and 11th, 1884*. Sunnyside, Orpington, Kent: George Allen, 1884. Bound in *The Works of John Ruskin*, ed. Edward Tyas Cook and Alexander Wedderburn, vol. 34. London: George Allen, 1908.

Ruskin, John. (Unconfirmed as photographer.) Daguerreotype. "West Facade, Santa Maria della Spina, Pisa," ca. 1846. *The Victorian Web: Literature, History, & Culture in the Age of Victoria*. http://www.victorianweb.org/painting/ruskin/daguerrotypes/6. html; accessed May 22, 2016.

Schaeffer, Kurtis R. *The University of Virginia's First Library: A Guide to Researching the First Library of the University of Virginia Using the 1828 Catalogue*. http://www.viseyes .org/library/ResearchGuide.pdf; accessed September 26, 2016.

Schaller, Dominik J., and Jürgen Zimmerer eds. *Journal of Genocide Research* 7 (4) (2005): 441–578.

Scheiner, Christoph. *Phantographice seu Ars delineandi res quaslibet per parallolegrammum lineare*. Roma: Grignani, 1631.

Scheunemann, Jürgen. *Top 10 Berlin*. London: DK Eyewitness Travel, 2014.

Scheunemann, Jürgen. "Top 10 Plundered Artifacts." *Time Magazine*, n.d.; http://content.time.com/time/specials/packages/article/0,28804,1883142_1883129_1883119,00 .html; accessed December 3, 2016.

Schneider, Peter. *Berlin Now: The City After the Wall*. Trans. Sophie Schlondorff. New York: Farrar, Straus and Giroux, 2014.

Schüller-Zwierlein, André. "Why Preserve? An Analysis of Preservation Discourses." *Preservation, Digital Technology & Culture* 44 (3) (2015): 98–122.

Schwartz, Hillel. *The Culture of the Copy: Striking Likenesses, Unreasonable Facsimiles*. New York: Zone Books, 1996.

Schwarz, Benjamin. "The Architect of the City." *The Atlantic*, March 2011. http://www.theatlantic.com/magazine/archive/2011/03/the-architect-of-the-city/308389; accessed February 5, 2016.

Scott, David A. "Conservation and Authenticity: Interactions and Enquiries." *Studies in Conservation* 60 (5) (2015): 291–305.

Scott Brown, Denise, Robert Venturi, and Steven Izenour. *Learning from Las Vegas*. Cambridge, MA: MIT Press, 1972. The book was revised in 1977, and a subtitle added: *The Forgotten Symbolism of Architectural Form*.

Semelin, Jacques. "Around the 'G' Word: From Raphael Lemkin's Definition to Current Memorial and Academic Controversies." *Genocide Studies and Prevention: An International Journal* 7 (1) (2012): 24–29.

Shaheen, Kareem, and Ian Black. "Beheaded Syrian Scholar Refused to Lead ISIS to Hidden Palmyra Antiquities." *The Guardian*, August 19, 2015. https://www.theguardian.com/world/2015/aug/18/isis-beheads-archaeologist-syria; accessed August 21, 2015.

Shelburne Museum. "Shelburne Museum." https://en.wikipedia.org/wiki/Shelburne_Museum; accessed September 20, 2016.

Siegel, Lee. "Twitter Can't Save You." *New York Times Book Review*, February 4, 2011, 14–15. (Review of Evgeny Morozov's *The Net Delusion: The Dark Side of Internet Freedom* [New York: PublicAffairs, 2011]). http://www.nytimes.com/2011/02/06/books/review/Siegel-t.html?pagewanted=all&_r=0; accessed October 12, 2016.

"Sisi Names Bomber of Cairo Coptic Christian Church," *Al Jazeera*, December 12, 2016.

"Situations under Investigation." Cour Pénale Internationale / International Criminal Court. https://www.icc-cpi.int/pages/situations.aspx?ln=en; accessed July 24, 2015.

Smith, Abby. "Valuing Preservation." *Library Trends* 56 (1) (Summer 2007): 4–25.

Sorokin, Vladimir. "Let the Past Collapse on Time!" *New York Review of Books*, May 8, 2014. http://www.nybooks.com/articles/archives/2014/may/08/let-the-past-collapse-on-time; accessed February 10, 2015.

Southern Regional Library Facility, University of California. "Chronology of Microfilm Developments, 1800–1900." *The History of Microfilm: 1839 to the Present*. https://www.google.com/webhp?tab=mw&ei=RDcWWNPnPInumQGEqqbYCA&ved=0EKkuCAUoAQ - q=microfilm+timeline; accessed October 29, 2016.

Spurr, Jeff. Personal communication, April 13, 2006.

Steele, Jonathan. "The Syrian Kurds Are Winning." *The New York Review of Books* 62 (19) (December 3, 2015). (Review of Michael M. Gunter. *Out of Nowhere: The Kurds of*

Syria in Peace and War. London: Hurst, 2015.) http://www.nybooks.com/articles/2015/12/03/syrian-kurds-are-winning; accessed August 19, 2016.

Stewart, Rob. *Sharkwater: The Photographs*. Toronto: Key Porter Books, 2007.

Stipe, Robert E., ed. *A Richer Heritage: Historic Preservation in the Twenty-first Century*. Chapel Hill: University of North Carolina Press, 2003.

Stolow, Nathan. *Controlled Environment for Works of Art in Transit*. Published for the International Centre for the Study of the Conservation of Cultural Property. London: Butterworths, 1966.

Stone, Peter. "War and Heritage: Using Inventories to Protect Cultural Property." *The Getty Conservation Newsletter* 28 (2) (Fall 2013). http://www.getty.edu/conservation/publications_resources/newsletters/28_2/war_heritage.html; accessed August 29, 2016.

Stossell, Sage. "The Architecture of Louis Sullivan: A Photo Gallery." *The Atlantic*, February 8, 2011. https://www.theatlantic.com/entertainment/archive/2011/02/the-architecture-of-louis-sullivan-a-photo-gallery/70108/#slide1; accessed February 5, 2016.

Stubbs, John H. *Time Honored: A Global View of Architectural Conservation: Parameters, Theory, & Evolution of an Ethos*. Hoboken, NJ: John Wiley & Sons, 2009.

Sturges, Paul. "Limits to Freedom of Expression? Considerations Arising from the Danish Cartoons Affair." *IFLA Journal* 32 (3) (2006): 181–188.

Sturges, Paul. "Limits to Freedom of Expression? The Problem of Basphemy." *IFLA Journal* 41 (2) (2015): 112–119.

Sudjic, Deyan. *Norman Foster: A Life in Architecture*. New York: Overlook Press, 2010.

Swartzberg, Susan. Holograph diary, August 16–27, 1991, copy in the possession of Michèle V. Cloonan.

Syria Untold. "About Syria Untold." http://www.syriauntold.com/en/about-syria-untold; accessed August 15, 2016.

Tanselle, G. Thomas. "A Rationale of Collecting." *Studies in Bibliography* 51 (1998): 1–20. https://forensicaesthetics.files.wordpress.com/2014/01/g-thomas-tanselle.pdf; accessed January 25, 2017.

Taylor, Ken, Archer St. Clair, and Nora J. Mitchell, eds. *Conserving Cultural Landscapes: Challenges and New Directions*. New York: Routledge, 2014.

Terkel, Studs. *Studs Terkel's Chicago*. New York: The New Press, 1986.

Tharoor, Ishaan. "The Bust of Nefertiti: Remembering Ancient Egypt's Most Famous Queen." *Time Magazine*, December 6, 2012. http://world.time.com/2012/12/06/the-bust-of-nefertiti-remembering-ancient-egypts-famous-queen; accessed December 3, 2016.

"30th Anniversary of Vietnam Wall." USA Snapshots. *USA Today*, October 31, 2012, 1.

Thompson, Janna. "Environment as Cultural Heritage." *Environmental Ethics* 22 (3) (Fall 2000): 241–258.

Towner, Lawrence W. *An Uncommon Collection of Uncommon Collections: The Newberry Library*. Chicago: The Newberry Library, 1970.

"'Treasures of Early Irish Art': An Exchange." *The New York Review of Books* 25 (12) (July 20, 1978): 47–48.

Turkle, Sherry, ed. *Evocative Objects: Things We Think With*. Cambridge, MA: MIT Press, 2007.

Tzu, Sun. *The Art of War*. Trans. Lionel Giles. Blacksburg, VA: Thrifty Books, 2009.

UNESCO. "Aleppo." (World Heritage site.) *Silk Road: Dialogue, Diversity & Development*. http://en.unesco.org/silkroad/content/aleppo; accessed August 15, 2016.

UNESCO. "Armed Conflict and Heritage: 1999 Second Protocol to the Hague Convention." http://www.unesco.org/new/en/culture/themes/armed-conflict-and-heritage/convention-and-protocols/1999-second-protocol; accessed September 18, 2017.

UNESCO. "Book of Kells: Documentary Heritage Submitted by Ireland and Recommended for Inclusion in the Memory of the World Register in 2011." *Memory of the World*. http://www.unesco.org/new/en/book-of-kells; accessed January 25, 2017.

UNESCO. Charter on the Preservation of Digital Heritage. Paris: UNESCO, October 15, 2003. http://portal.unesco.org/en/ev.php-URL_ID=17721&URL_DO=DO_TOPIC&URL_SECTION=201.html; accessed December 21, 2016.

UNESCO. Convention for the Safeguarding of the Intangible Cultural Heritage. Paris: UNESCO, October 17, 2003; http://portal.unesco.org/en/ev.php-URL_ID=17716&URL_DO=DO_TOPIC&URL_SECTION=201.html; accessed December 21, 2016.

UNESCO. "Expert Working Group Releases Recommendations for Safeguarding Bamiyan." Paris: UNESCO World Heritage Convention, April 27, 2011. whc.unesco.org; accessed December 1, 2016.

UNESCO. "UNESCO: Building Peace in the Minds of Men and Women: Introducing UNESCO: What We Are." http://www.unesco.org/new/en/unesco/about-us/who-we-are/introducing-unesco; accessed July 22, 2015.

UNESCO. "UNESCO: Building Peace in the Minds of Men and Women: The Organization's History." http://www.unesco.org/new/en/unesco/about-us/who-we-are/history; accessed July 2, 2015.

UNESCO. *UNESCO/PERSIST (Platform to Enhance the Sustainability of the Information Society Transglobally) Guidelines for the Selection of Digital Heritage for Long-Term Preservation*. UNESCO/PERSIST Content Task Force, March 2016. https://www.unesco.nl

/sites/default/files/uploads/Comm_Info/persistcontentguidelinesfinal1march2016.pdf; accessed September 18, 2017.

UNESCO. "What Is Intangible Cultural Heritage?" https://ich.unesco.org/en/what-is-intangible-heritage-00003; accessed October 5, 2017.

United Nations Convention on the Prevention and Punishment of the Crime of Genocide (1948). http://www.hrweb.org/legal/genocide.html; accessed July 2, 2015.

United Nations Declaration on the Rights of Indigenous Peoples (2007). http://www.un.org/esa/socdev/unpfii/documents/DRIPS_en.pdf; accessed July 18, 2015.

United Nations Permanent Forum on Indigenous Issues. "Who Are Indigenous Peoples?" Factsheet. *Indigenous Peoples, Indigenous Voices*. http://www.un.org/esa/socdev/unpfii/documents/5session_factsheet1.pdf; accessed January 28, 2017.

United Nations Security Council. "Letter Dated 24 May 1994 from the Secretary-General to the President of the Security Council." http://www.icty.org/x/file/About/OTP/un_commission_of_experts_report1994_en.pdf; accessed July 18, 2015.

University of California Museum of Paleontology. "Georges-Louis Leclerc, Comte de Buffon (1707–1788)." http://www.ucmp.berkeley.edu/history/buffon2.html; accessed July 21, 2016.

U.S. Department of State, Office of the Historian. "The Kellogg-Briand Pact, 1928." N.d. https://history.state.gov/milestones/1921-1936/kellogg; accessed June 30, 2015.

"The USSR's 1948 Instructions for the Identification, Registration, Maintenance, and Restoration of Architectural Monuments under State Protection." Trans. Richard Anderson. "Special Issue on the Preservation of Soviet Heritage," ed. Jean-Louis Cohen and Barry Bergdoll. *Future Anterior: Journal of Historic Preservation, History, Theory, and Criticism* 5 (1) (Summer 2008): 64–72.

van Dam, Nikolaos. *The Struggle for Power in Syria: Politics and Society under Asad and the Ba'th Party*. London: I. B. Tauris, 1997.

Varlamoff, Marie-Thérèse, and George MacKenzie. "Archives in Times of War: The Role of IFLA and ICA within ICBS (International Committee of the Blue Shield)." In *A Reader in Preservation and Conservation, IFLA Publications 91*, ed. Ralph W. Manning and Virginie Kremp, 149–157. München: K. G. Saur, 2000; reprinted 2013. An incomplete online version is at https://books.google.com/books?id=-XchAAAAQBAJ&pg=PA149&lpg=PA149&dq=archives+in+times+of+war+varlamoff&source=bl&ots=Th6sYFPFfj&sig=9kWmtegQI-Qr1293_zRxkFCOOSs&hl=en&sa=X&ved=0ahUKEwjdxq6-2eLRAhVCj1QKHeFqAE4Q6AEIHDAA - v=onepage&q=archives in t; accessed January 27, 2017.

"Views." Letters by Thomas B. Stauffer and Le Corbusier on the Garrick Theater in Chicago, published in *Progressive Architecture* 42 (6) (June 1961): 208.

Walsh, Caroline. "Art Treasures to Be Shown in US." *Irish Times*, June 30, 1977, 9.

Weever, John. *Ancient Funeral Monuments within the United Monarchie of Great Britain.* London: Thomas Harper, 1631.

Welchman, Jennifer. "The Virtues of Stewardship." *Environmental Ethics* 21 (4) (Winter 1999): 411–423.

Wikipedia. "Carbon Paper." https://en.wikipedia.org/wiki/Carbon_paper; accessed October 29, 2016.

Wikipedia. "Fallen Monument Park." http://en.wikipedia.org/wiki/Fallen_Monu ment_Park; accessed February 15, 2011.

Wikipedia [German language]. "Hiroshimastrasse." https://de.wikipedia.org/wiki /Hiroshimastraße; accessed January 29, 2017.

Wikipedia. "Human Rights." https://en.wikipedia.org/wiki/Human_rights; accessed June 30, 2015.

Wikipedia. "Leaning Tower of Pisa." https://en.wikipedia.org/wiki/Leaning_Tower _of_Pisa; accessed July 10, 2016.

Wikipedia. "Richard Nickel." https://en.wikipedia.org/wiki/Richard_Nickel; accessed February 8, 2016.

Williams, Roddy, Director of Operations, NAMES Project/AIDS Memorial Quilt; email communication, March 26, 2016.

Wordsworth, Andrew. "Have Italy's Art Restorers Cleaned up Their Act?" *The Independent*, June 20, 2000. http://www.independent.co.uk/incoming/have-italys-art -restorers-cleaned-up-their-act-5370823.html; accessed January 28, 2017.

Wright, David H. "Correspondence." *The New York Review of Books* 25 (12) (July 20, 1978): 47–48.

Wright, David H. "Shortchanged at the Met." *The New York Review of Books* 25 (7) (May 4, 1978): 32–34.

WTTW. *Chicago Stories.* http://interactive.wttw.com/a/chicago-stories; accessed January 28, 2017.

WTTW. "The Richard Nickel Story." *Chicago Stories*, 2002. http://interactive.wttw. com/a/chicago-stories-richard-nickel-story; accessed January 28, 2017. The full video is available at http://video.wttw.com/video/2365335740/; accessed January 28, 2017.

Yassin-Kassab, Robin, and Leila Al-Shami. *Burning Country: Syrians in Revolution and War.* London: Pluto Press, 2016.

Yates, Frances A. *The Art of Memory.* Chicago: University of Chicago Press, 1966.

Yildiz, Kerim. *The Kurds in Syria: The Forgotten People.* London: Pluto Press, 2005.

Index

Note: italic page numbers refer to figures.